SpringerBriefs in Computer Science

AF294400

SpringerBriefs present concise summaries of cutting-edge research and practical applications across a wide spectrum of fields. Featuring compact volumes of 50 to 125 pages, the series covers a range of content from professional to academic.

Typical topics might include:

- A timely report of state-of-the art analytical techniques
- A bridge between new research results, as published in journal articles, and a contextual literature review
- A snapshot of a hot or emerging topic
- An in-depth case study or clinical example
- A presentation of core concepts that students must understand in order to make independent contributions

Briefs allow authors to present their ideas and readers to absorb them with minimal time investment. Briefs will be published as part of Springer's eBook collection, with millions of users worldwide. In addition, Briefs will be available for individual print and electronic purchase. Briefs are characterized by fast, global electronic dissemination, standard publishing contracts, easy-to-use manuscript preparation and formatting guidelines, and expedited production schedules. We aim for publication 8–12 weeks after acceptance. Both solicited and unsolicited manuscripts are considered for publication in this series.

**Indexing: This series is indexed in Scopus, Ei-Compendex, and zbMATH **

Rubén Ballester • Carles Casacuberta •
Sergio Escalera

Topological Data Analysis for Neural Networks

 Springer

Rubén Ballester
Mathematics and Computer Science
Universitat de Barcelona
Barcelona, Spain

Carles Casacuberta
Mathematics and Computer Science
Universitat de Barcelona
Barcelona, Spain

Sergio Escalera
Mathematics and Computer Science
Universitat de Barcelona
Barcelona, Spain

ISSN 2191-5768   ISSN 2191-5776   (electronic)
SpringerBriefs in Computer Science
ISBN 978-3-032-08282-4   ISBN 978-3-032-08283-1   (eBook)
https://doi.org/10.1007/978-3-032-08283-1

© The Editor(s) (if applicable) and The Author(s), under exclusive license to Springer Nature Switzerland AG 2026

This work is subject to copyright. All rights are solely and exclusively licensed by the Publisher, whether the whole or part of the material is concerned, specifically the rights of translation, reprinting, reuse of illustrations, recitation, broadcasting, reproduction on microfilms or in any other physical way, and transmission or information storage and retrieval, electronic adaptation, computer software, or by similar or dissimilar methodology now known or hereafter developed.
The use of general descriptive names, registered names, trademarks, service marks, etc. in this publication does not imply, even in the absence of a specific statement, that such names are exempt from the relevant protective laws and regulations and therefore free for general use.
The publisher, the authors and the editors are safe to assume that the advice and information in this book are believed to be true and accurate at the date of publication. Neither the publisher nor the authors or the editors give a warranty, expressed or implied, with respect to the material contained herein or for any errors or omissions that may have been made. The publisher remains neutral with regard to jurisdictional claims in published maps and institutional affiliations.

This Springer imprint is published by the registered company Springer Nature Switzerland AG
The registered company address is: Gewerbestrasse 11, 6330 Cham, Switzerland

If disposing of this product, please recycle the paper.

Preface

Since the decade of the 2010s, topological data analysis (TDA) has permeated almost all scientific disciplines as a toolwork for studying datasets where shape features of data provide relevant information. The impact of TDA on machine learning, and deep learning (DL) in particular, has been especially significant, as evidenced by its growing influence and increasing number of related publications. Some of the applications of TDA to DL encompass the development of topology-aware machine learning pipelines, including novel neural network layers and architectures, as well as the examination of crucial neural network properties such as their generalization capacity through a topological lens. The successful application of TDA to DL has encouraged many researchers to explore this intersection over the last decade, generating a rich corpus of literature with diverse and valuable insights for machine learning practitioners.

The present monograph aims to provide a comprehensive overview of the growing body of literature that applies TDA techniques to analyze the different components of DL pipelines. Chapter 2 provides a brief yet detailed presentation of deep learning from a mathematical perspective, and Chap. 3 summarizes the main aspects of TDA theory and practice.

The remaining chapters discuss several ways in which topology is applied to the structure and dynamics of deep networks, primarily focusing on the content of 43 articles, including three papers published by the authors. Of the selected articles, only 11 were published before 2020, while 13 appeared in 2023 and 2024. This steady increase highlights the rapid development of the field and makes it evident that surveying applications of TDA to neural networks is not an easy task. First, because valuable new articles, possibly even breakthroughs, may emerge during or after the publication of this book. Second, despite our intention to be as exhaustive as possible, such a survey will inevitably omit important contributions that are not directly related to deep learning algorithms.

Nevertheless, this monograph aspires to serve as a resource for researchers at various stages of their careers. For those embarking on studies in topological deep learning, we provide sufficient background to build a solid foundation in the field. Simultaneously, we offer a thorough overview of the state of the art as of the volume's completion, which we believe will benefit even experienced researchers in navigating the extensive body of work developed during the last years.

We hope that this contribution will not only enhance understanding of the broad scope of TDA but also inspire future progress in the field, fostering continued innovation at the intersection of topology and deep learning.

Barcelona, Spain
July 2025

Rubén Ballester
Carles Casacuberta
Sergio Escalera

Acknowledgements This work was supported by the State Research Agency (AEI) of the Ministry of Science and Innovation of the Government of Spain, through research projects PID2019-105093GB-I00, PID2020-117971GB-C22, and PID2022-136436NB-I00; the Ministry of Universities through contract FPU21/00968; the Agency for Management of University and Research Grants (AGAUR) of the Government of Catalonia through grant 2021 SGR 00697; and the Catalan Institution for Research and Advanced Studies (ICREA) under the ICREA Acadèmia program.

Competing Interests The authors have no competing interests to declare that are relevant to the content of this manuscript.

Contents

Acronyms

AI	Artificial intelligence
BERT	Bidirectional encoder representations from transformers
CE	Categorical cross entropy loss
CIFAR	Canadian Institute for Advanced Research
CNN	Convolutional neural network
DBSCAN	Density-based spatial clustering of applications with noise
DCGAN	Deep convolutional generative adversarial network
DFC	Directed flag complex
DL	Deep learning
FCFNN	Fully connected feedforward neural network
FID	Fréchet inception distance
GAN	Generative adversarial network
GTDA	Graph-based topological data analysis
HVS	Human visual system
KDE	Kernel density estimator
LLM	Large language model
LVR	Labeled Vietoris–Rips
KNN	k-Nearest neighbors
ML	Machine learning
MNIST	Modified National Institute of Standards and Technology
MRLT	Mean relative living time
PCA	Principal component analysis
PGDL	Predicting Generalization in Deep Learning
ReLU	Rectified linear unit
RL	Reinforcement learning
RLT	Relative living time
RTD	Representation topology divergence
SSCC	Signed sequence cubical complex
SVHN	Street view house numbers
SVM	Support vector machine
TDA	Topological data analysis

TDL	Topological deep learning
TP	Total persistence
UCI	University of California, Irvine
VGG	Visual Geometry Group
VR	Vietoris–Rips
WGAN	Wasserstein generative adversarial network

Chapter 1
Introduction

Abstract This monograph presents a comprehensive account of methods from topological data analysis, such as persistent homology and Mapper graphs, applied to the study of neural network structure and dynamics. We describe various strategies for extracting topological information from data, and examine how such information can be used to analyze properties of neural networks, including their generalization capacity and expressivity. We also discuss practical implications of the usage of topological data analysis in deep learning applications, with a focus on adversarial detection and model selection. The monograph concludes with a discussion of current challenges and potential future developments in the field.

1.1 Overview

Over the past few years, deep learning has consolidated its position as the most successful branch of artificial intelligence. With the continuous growth in computational capacity, neural networks have expanded in size and complexity, enabling them to effectively tackle progressively difficult problems. However, their increased capacity has made it more challenging to comprehend essential properties of the networks such as their interpretability, generalization ability, or suitability for specific problems. From both theoretical and practical standpoints, this is undesirable, especially in critical contexts where AI decisions could lead to catastrophic consequences, such as medical diagnosis (Yang et al. 2021) or autonomous driving (Wäschle et al. 2022), among others.

Topological data analysis (TDA) has emerged as a subfield of algebraic topology that offers a framework for gaining insights into the *shape* of data in a broad sense. Topological data analysis has found application across a wide array of experimental science disciplines, including biomedicine (Skaf and Laubenbacher 2022), finance (Gidea and Katz 2018), and numerous others. One of its most prolific areas of application is machine learning, particularly in the domain of deep learning. A basic introduction to topological machine learning can be found in Hensel et al. (2021). Topological data analysis, specifically homology, persistent homology and

© The Author(s), under exclusive license to Springer Nature Switzerland AG 2026
R. Ballester et al., *Topological Data Analysis for Neural Networks*, SpringerBriefs in Computer Science, https://doi.org/10.1007/978-3-032-08283-1_1

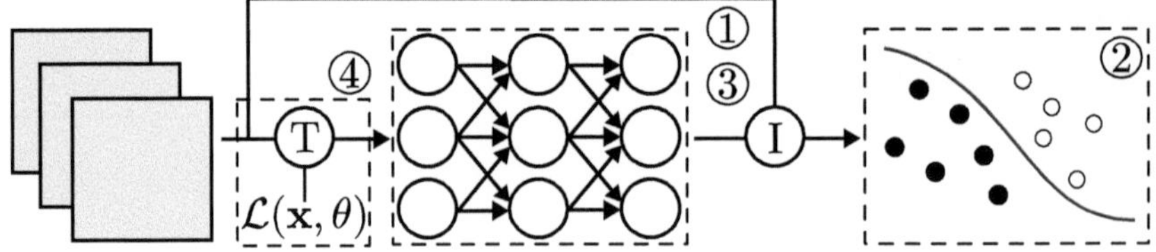

Fig. 1.1 Diagram showing the usual lifecycle of a neural network $\mathcal{N}$. First, an architecture $a(\mathcal{N})$ is selected based on the task to be solved. This architecture is independent of the learned parameters $\theta(\mathcal{N})$ or the specific input data used to train or test the network. Second, the architecture is trained (T) using a specific training algorithm, which generally minimizes the empirical risk of a loss function $\mathcal{L}$ evaluated on the training dataset. Once the network is trained, inference (I) is performed with data sharing the same distribution as the training data. For trained networks, input and output spaces gather interesting structures, such as decision regions and boundaries for classification problems or latent spaces for generative models. Each dashed box is related to one of the categories in which TDA has been used, namely: (1) Architecture; (2) Input and output spaces; (3) Internal representations and activations; (4) Training dynamics and loss functions. The rightmost box depicts decision regions and decision boundary of the output space

Mapper, has been used to analyze various aspects of neural networks. Broadly, these aspects can be categorized in the following four groups:

1. Architecture;
2. Input and output spaces;
3. Internal representations and activations;
4. Training dynamics and loss functions.

Figure 1.1 visually delineates these four categories. The first category consists of aspects related to unweighted graphs associated with neural network architectures and properties of these graphs, such as depth, layer widths, and graph topology, among others. The second category encompasses input and output spaces, including decision regions and boundaries for classification problems, as well as latent spaces for generative models. The third category, which currently holds the largest number of contributions in the literature, covers aspects related to hidden and output neurons in a broad sense. Lastly, the fourth category comprises neural network training procedures, including loss functions.

Many components studied within the preceding categories play a pivotal role in understanding some of the fundamental traits of deep learning, such as interpretability or the generalization capacity of neural networks. Moreover, these components inherently exhibit geometrical and topological characteristics, rendering them exceptionally suitable for the application of topological analysis methods.

1.2 Outline

In this work, we offer a comprehensive overview of applications of TDA in analyzing neural networks across the aforementioned four categories. However, we have omitted many relevant works related to the general use of TDA in deep learning. For example, we omit work on the branch of topological deep learning (TDL), which involves developing neural networks tailored for specific topological data structures. A recent survey on topological deep learning is available in Papillon et al. (2023). It is important to remark that the term *topological deep learning* has been used to describe two distinct yet complementary areas within deep learning. The first involves, generically, the application of topological methods to standard deep learning, as discussed in the preface and in this introduction. The second focuses on developing neural network architectures designed for topological domains, such as simplicial or cellular complexes. Unless otherwise stated, we use the term topological deep learning to refer to the latter approach.

We have also omitted the majority of applications using TDA to construct loss functions, since our focus is primarily on the analysis of neural networks rather than their enhancement through the imposition of specific topological structures on data, unrelated to the particular machine learning algorithm used for the task.

Given the substantial volume of papers analyzed, our discussion is limited to peer-reviewed publications, with occasional exceptions made for relevant and credible sources. We apologize for any unintentional omissions.

In Part I we introduce notation, definitions, and basic results of TDA and deep learning needed to follow the rest of the book. Part II constitutes the core content of our survey, structured into the four aforementioned categories. Section 6.2.1 contains a summary of the authors' contributions through a study of activations in the complete neural network graph, aimed at analyzing the generalization capabilities of deep networks. Chapter 8 addresses limitations, challenges, and potential future directions concerning the application of TDA in deep learning up to the time of publication of these notes.

This book aims to be self-contained and approachable for readers unfamiliar with topological data analysis or deep learning. However, it is advisable that readers have a background in at least one of these areas. For mathematicians interested in how topology can be applied to study modern AI and deep learning systems, we recommend reading the first chapter of the book by Grohs and Kutyniok (2022), that provides a concise introduction to deep learning from a mathematical viewpoint. For machine learning scientists curious about how advanced topological methods can enhance the understanding of deep learning systems, we recommend the survey on topological machine learning by Hensel et al. (2021), where TDA is introduced in a friendly manner within the scope of machine learning.

Preliminary definitions and results for the deep learning part are primarily drawn from the first chapter of the book by Grohs and Kutyniok (2022). Regarding the topological data analysis part, some of the content is sourced from the book by Edelsbrunner and Harer (2022) and the survey conducted by Hensel et al. (2021).

1.3   Notation

We denote by $\mathbb{N}$, $\mathbb{Z}$, $\mathbb{R}$, and $\bar{\mathbb{R}}$ respectively the set of natural numbers (including zero), the ring of integers, the field of real numbers, and the extended real line with a point at infinity. For $n \geq 1$, we abbreviate $\{1, \ldots, n\}$ as $[n]$.

Given a set S, its power set, i.e., the set of its subsets, is denoted by $\mathcal{P}(S)$. The indicator function on a set A is written as $\mathbb{1}_A$ and it is defined as $\mathbb{1}_A(x) = 1$ if $x \in A$ and $\mathbb{1}_A(x) = 0$ if $x \notin A$. The Kronecker delta $\delta_{i,j}$ is defined as $\delta_{i,j} = 1$ if $i = j$ and $\delta_{i,j} = 0$ if $i \neq j$. The cardinality of a finite set S is denoted by $|S|$.

The set of matrices of size $m \times n$ with coefficients in a ring R (usually a field) is denoted by $M_{m,n}(R)$. The (i, j) entry of a matrix M is written as $M_{i,j}$. For a square matrix M, its trace is denoted by $\mathrm{tr}(M)$. Given a vector $v \in \mathbb{R}^d$, we denote its i-th component by v_i.

In this work, graphs are simple (i.e., with at most one edge between any two given vertices and without loops at any vertex) and undirected, unless otherwise specified. For a graph G, we denote by $V(G)$ and $E(G)$ the sets of vertices and edges of G, respectively. By a *weighted graph* we refer to an undirected graph G whose vertices and edges are weighted by functions $w_V \colon V(G) \to \mathbb{R}$ and $w_E \colon E(G) \to \mathbb{R}$, and we use the notation (G, w_V, w_E) for a weighted graph. A bipartite graph is denoted by $G = (V_1, V_2, E)$, where V_1 and V_2 are disjoint sets of vertices.

We denote by $\mathcal{M}(\mathcal{X}, \mathcal{Y})$ the set of all measurable functions from $\mathcal{X}$ to $\mathcal{Y}$, where $\mathcal{X}$ and $\mathcal{Y}$ are measure spaces. Given a function f, we denote its domain and image as $\mathrm{Dom}(f)$ and $\mathrm{Im}(f)$, respectively, and we denote its graph by

$$G(f) = \{(x, f(x)) : x \in \mathrm{Dom}(f)\}.$$

The *support* of a function $f \colon X \to \mathbb{R}$, denoted $\mathrm{supp}(f)$, is the set of elements of X that are not sent to zero, or most commonly the closure of this set if X is a topological space. Similarly, for a probability distribution $\mathbb{P}$ associated with a random variable X, the support $\mathrm{supp}(\mathbb{P})$ is the smallest closed set S of real numbers such that $\mathbb{P}(X \in S) = 1$. We also call support of $\mathbb{P}$ the set of elements of the sample space mapped to $\mathrm{supp}(\mathbb{P})$ by X. The letter $\mathbb{E}$ is used to denote expectation of a random variable.

References

Herbert Edelsbrunner and John Write L. Harer. *Computational Topology: An Introduction.* American Mathematical Society, 2022.

Marian Gidea and Yuri Katz. Topological data analysis of financial time series: Landscapes of crashes. *Physica A: Statistical Mechanics and its Applications*, 491:820–834, 2018. https://doi.org/10.1016/j.physa.2017.09.028. URL https://www.sciencedirect.com/science/article/pii/S0378437117309202.

Philipp Grohs and Gitta Kutyniok. *Mathematical Aspects of Deep Learning.* Cambridge University Press, 2022. https://doi.org/10.1017/9781009025096.

Felix Hensel, Michael Moor, and Bastian Rieck. A survey of topological machine learning methods. *Frontiers in Artificial Intelligence*, 4, 2021. ISSN 2624-8212. https://doi.org/10.3389/frai.2021.681108. URL https://www.frontiersin.org/articles/10.3389/frai.2021.681108.

Mathilde Papillon, Sophia Sanborn, Mustafa Hajij, and Nina Miolane. Architectures of topological deep learning: A survey on topological neural networks, 2023. URL https://arxiv.org/abs/2304.10031.

Yara Skaf and Reinhard Laubenbacher. Topological data analysis in biomedicine: A review. *Journal of Biomedical Informatics*, 130:104082, 2022. ISSN 1532-0464. https://doi.org/10.1016/j.jbi.2022.104082. URL https://www.sciencedirect.com/science/article/pii/S1532046422000983.

Moritz Wäschle, Florian Thaler, Axel Berres, Florian Pölzlbauer, and Albert Albers. A review on AI Safety in highly automated driving. *Frontiers in Artificial Intelligence*, 5, 2022. ISSN 2624-8212. https://doi.org/10.3389/frai.2022.952773. URL https://www.frontiersin.org/articles/10.3389/frai.2022.952773.

Sijie Yang, Fei Zhu, Xinghong Ling, Quan Liu, and Peiyao Zhao. Intelligent health care: Applications of deep learning in computational medicine. *Frontiers in Genetics*, 12, 2021. ISSN 1664-8021. https://doi.org/10.3389/fgene.2021.607471. URL https://www.frontiersin.org/articles/10.3389/fgene.2021.607471.

Part I
Fundamentals

Chapter 2
Deep Learning

Abstract Deep learning is a subfield of machine learning that deals with multilayer neural networks to solve a variety of computational tasks. Such tasks include, but are not limited to, classification, regression, and synthetic data generation. These three kinds of tasks are central in this book.

2.1 Computational Tasks

Classification and regression are particular examples of *prediction* tasks. In prediction tasks, datasets can be formally viewed as subsets of the Cartesian product of two measure spaces $\mathcal{X}$ and $\mathcal{Y}$, where elements of $\mathcal{Y}$ are considered *labels* of elements in $\mathcal{X}$. Such data are distributed according to a distribution $\mathbb{P}_{(X,Y)}$ with marginals $\mathbb{P}_X$ and $\mathbb{P}_Y$ for the restrictions to $\mathcal{X}$ and $\mathcal{Y}$, respectively.

In prediction tasks, one looks for a measurable function $f : \mathcal{X} \to \mathcal{Y}$ that, given a value $x \in \mathcal{X}$, outputs a label prediction $y \in \mathcal{Y}$ according to some quality criterion. To precisely define the concept of quality, a *loss function*

$$\mathcal{L} : \mathcal{M}(\mathcal{X}, \mathcal{Y}) \times \mathcal{X} \times \mathcal{Y} \longrightarrow \mathbb{R}$$

is typically employed, where $\mathcal{M}(\mathcal{X}, \mathcal{Y})$ denotes the space of measurable functions from $\mathcal{X}$ to $\mathcal{Y}$. A loss function indicates the degree of inaccuracy in predictions relative to a data point $(x, y) \in \mathcal{X} \times \mathcal{Y}$ sampled from the data distribution —where a lower value signifies greater quality.

Thus, the objective is to find a function f that minimizes the expected loss $\mathbb{E}\left[\mathcal{L}(f, x, y)\right]$, called *risk* of f and denoted by $\mathcal{R}(f)$. Actually, a minimum of this expectation might not exist, in which case one looks for a function f arbitrarily close to attaining the infimum of $\mathbb{E}\left[\mathcal{L}(f, x, y)\right]$.

Due to the hardness of finding such an f (if it exists) in the space $\mathcal{M}(\mathcal{X}, \mathcal{Y})$, the problem is often reduced to finding a function f whose risk is arbitrarily near to the infimum of risks in a smaller set of functions, called *hypothesis set*.

A *neural network* is a parameterized function $\mathcal{N}_\theta : \mathcal{X} \to \mathcal{Y}$ in a given hypothesis set, which is expected to possess desirable properties such as being a universal

© The Author(s), under exclusive license to Springer Nature Switzerland AG 2026
R. Ballester et al., *Topological Data Analysis for Neural Networks*, SpringerBriefs in Computer Science, https://doi.org/10.1007/978-3-032-08283-1_2

approximator of a large class of functions (Grohs and Kutyniok 2022, Chapter 3) or being able to overcome the curse of dimensionality (Bronstein et al. 2021, Chapter 2.2).

One of the main difficulties in finding an optimal function f according to the risk minimization criterion is that the data distribution $\mathbb{P}_{(X,Y)}$ is often unknown, and we only have access to a finite sample $\mathcal{D} = \{(x_i, y_i)\}_{i=1}^{m}$ from $\mathcal{X} \times \mathcal{Y}$, that we assume independent and identically distributed, with sampling distribution $(\mathbb{P}_{(X,Y)})^m$. Therefore, instead of trying to minimize directly the risk $\mathcal{R}(f)$, one tries to minimize the *empirical risk* for a subset $\mathcal{D}_{\text{train}} \subseteq \mathcal{D}$, called *training dataset*, given by

$$\widehat{\mathcal{R}}_{\mathcal{D}_{\text{train}}}(f) = \frac{1}{m} \sum_{i=1}^{m} \mathcal{L}(f, x_i, y_i), \tag{2.1}$$

with the hope that minimizing the empirical risk also reduces the real risk.

In some situations, it is convenient to minimize the empirical risk of another related loss $\mathcal{L}^{\text{surr}}$ that is not the original loss function $\mathcal{L}$ and that can depend on a surrogate function f^{surr} depending on f, instead of using f directly. This happens, for example, when we use minimization algorithms that rely on the differentiability of the loss function, yet our function $\mathcal{L}$ is not differentiable or its gradient is zero.

In deep learning, given data distributed according to $\mathbb{P}_{(X,Y)}$, one formally sees the process of obtaining a neural network as a map

$$\mathcal{A}: \bigcup_{m \in \mathbb{N}} (\mathcal{X} \times \mathcal{Y})^m \longrightarrow \mathcal{F}, \tag{2.2}$$

where $\mathcal{F}$ is some hypothesis set that takes as input an ordered set $\mathcal{D} = \{(x_i, y_i)\}_{i=1}^{m}$ and provides a neural network $\mathcal{N} \in \mathcal{F}$, that is usually obtained by trying to minimize the empirical risk for the original or surrogate loss on a subset $\mathcal{D}_{\text{train}} \subseteq \mathcal{D}$. A map such as (2.2) is called a *training algorithm*. Training algorithms are generally iterative minimization methods based on gradient descent.

Regression and classification are the two modalities studied for prediction tasks in this work. In both cases, $\mathcal{X}$ is a Euclidean space $\mathbb{R}^d$ for some $d \geq 1$. We say that a prediction task is a *regression task* if $\mathcal{Y} = \mathbb{R}$, and we call it a *classification task* if $\mathcal{Y}$ is a finite set. In the latter case we assume, without loss of generality, that $\mathcal{Y} = \{1, \ldots, k\}$ with $k \geq 2$, and that $\text{supp}(\mathbb{P}_Y) = \mathcal{Y}$. If $k = 2$, then we call it a *binary* classification task. For a broader introduction to machine learning tasks, we refer the reader to Goodfellow et al. (2016, Section 5.1.1).

Generative tasks are not as straightforward to define as prediction tasks, and formal definitions can vary depending on the problem or the solving method. In generative tasks, the purpose is to generate synthetic data that are indistinguishable from source data. We assume that source data live in an ambient space $\mathcal{X}$ and follow a distribution $\mathbb{P}_X$. Some models explicitly approximate the probability distribution $\mathbb{P}_X$, allowing to directly sample from it, while other models learn a mechanism to generate new examples without explicitly describing the data distribution. We refer to Prince (2023, Section 1.2.1) for a detailed explanation of generative models.

Reinforcement learning tasks are also considered in one of the articles analyzed in this monograph. In reinforcement learning, an *agent* (e.g., a robot) interacts with an environment over discrete time steps aiming to optimize a given notion of *rewards* generated by its actions. In deep reinforcement learning, neural networks are used to control the behaviour of the agent in various ways. These methods include approximating the value for the agent of being in a specific state or taking a particular action within a state (Mnih et al. 2015), as well as extracting and transforming relevant information from the agent's environment (Le Lan 2023), among other techniques. Similar to the previous tasks, agents typically undergo a training phase, where they learn to behave effectively in the environment, and an inference phase, where they interact with the environment without learning from it. For a comprehensive introduction to reinforcement learning, we refer the reader to François-Lavet et al. (2018).

2.2 Fully Connected Feedforward Neural Networks

In this book we mainly consider *fully connected feedforward neural networks* (FCFNNs). Such networks have a basic structure in which most of the current models can be expressed. A FCFNN is a parameterized function $\mathcal{N}$ consisting of the following elements.

(i) An *architecture* $a(\mathcal{N}) = (N, \varphi)$, where $N = (N_0, \ldots, N_L) \in \mathbb{N}^{L+1}$ and

$$\varphi = \left(\varphi^{(l)}\right)_{l=1}^{L}, \qquad \varphi^{(l)} : \mathbb{R}^{N_l} \to \mathbb{R}^{N_l}$$

is a tuple of non-linear differentiable functions. Here differentiability is meant except perhaps on a set of zero measure. The number L counts the *layers* of the network, while N_0, N_L, and N_l for $l \in \{1, \ldots, L - 1\}$ are the numbers of neurons of the input, output, and l-th hidden layers, respectively, and $\varphi^{(l)}$ is the *activation function* of layer l. Usually, the activation functions $\varphi^{(l)}$ are chosen to be functions which apply a surrogate function $\varphi_e^{(l)} : \mathbb{R} \to \mathbb{R}$ componentwise, like the *rectified linear unit* (ReLU) function $\mathrm{ReLU}(x) = \max\{0, x\}$. If not specified otherwise, we assume that activation functions are calculated by applying surrogate functions componentwise.

(ii) A set of parameters

$$\theta(\mathcal{N}) = \left(\theta^{(l)}\right)_{l=1}^{L}, \qquad \theta^{(l)} = \left(W^{(l)}, b^{(l)}\right) \in \mathbb{R}^{N_l \times N_{l-1}} \times \mathbb{R}^{N_l}$$

for $l \in \{1, \ldots, L\}$, where entries in the matrix $W^{(l)}$ are called *weights* and those in the vector $b^{(l)}$ are called *biases*.

A FCFNN induces a directed graph and a function depending on it. The directed graph defined by the architecture, denoted by $G(\mathcal{N})$, is given by $L + 1$ disjoint ordered sets $V_l = \{v_l^1, \ldots, v_l^{N_l}\}$, $l = 0, \ldots, L$, with the property that every vertex in V_l has an edge pointing to it from all the vertices in V_{l-1}, for $l \in \{1, \ldots, L\}$. The vertices of this graph are called *neurons*.

The function $\phi_{\mathcal{N}} \colon \mathbb{R}^{N_0} \to \mathbb{R}^{N_L}$ defined on top of the directed graph, which represents the FCFNN, is given by the recursive formula

$$\phi_{\mathcal{N}}^{(0)}(x) = x,$$

$$\phi_{\mathcal{N}}^{(l)}(x) = \varphi^{(l)}\left(W^{(l)}\phi_{\mathcal{N}}^{(l-1)}(x) + b^{(l)}\right) \quad \text{for } l \in \{1, \ldots, L\}, \tag{2.3}$$

$$\phi_{\mathcal{N}}(x) = \phi_{\mathcal{N}}^{(L)}(x).$$

Given an input vector x, we say that the values $\phi_{\mathcal{N}}^{(l)}(x)_i$ are *activations* of the corresponding vertices v_l^i. A visual example of a FCFNN is shown in Fig. 2.1.

Given an architecture $a = (N, \varphi)$ for a FCFNN and a subset of *admissible* parameters $\Theta \subseteq \prod_{l=1}^{L} \mathbb{R}^{N_l \times N_{l-1}} \times \mathbb{R}^{N_l}$, we denote the hypothesis set of all the neural networks represented by the architecture a with parameters in Θ as

$$\mathcal{F}_{a,\Theta} = \{\mathcal{N} : a(\mathcal{N}) = a, \ \theta(\mathcal{N}) \in \Theta\}.$$

If Θ is the space of all possible parameters, we simply write $\mathcal{F}_a$. Also, when it is clear that we speak about neural networks in a specific hypothesis set $\mathcal{F}_{a,\Theta}$, we denote the neural network with architecture a and parameters $\theta \in \Theta$ as $\mathcal{N}_\theta$.

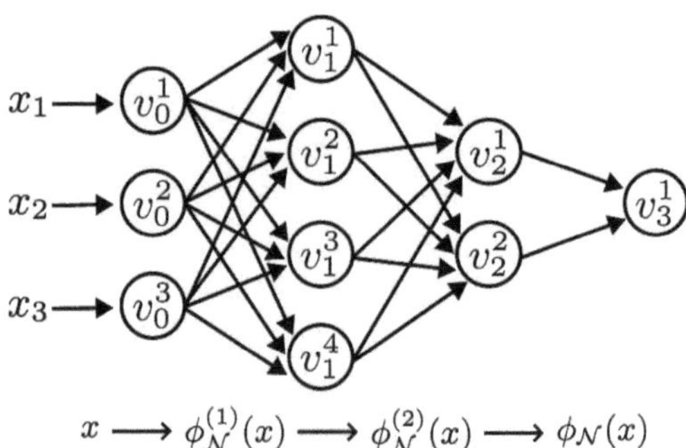

Fig. 2.1 A graphical representation of a fully connected feedforward neural network $\mathcal{N}$ with $L = 3$ and $N = (3, 4, 2, 1)$. The input values x_i are associated with the vertices v_0^i of the input layer and transformed sequentially by a set of maps. In this representation, each edge indicates that the value of the source vertex is used for the computation of the value of the target vertex. Values for vertices are computed sequentially from the first layer to the last. A transformation from layer $l - 1$ to layer l is a composite of an affine transformation and the activation function $\varphi^{(l)}$ applied componentwise

Fixing a dataset $\mathcal{D}$, an architecture a, and a set of possible parameters Θ, the composition of the empirical risk with the neural network functions in $\mathcal{F}_{a,\Theta}$ can be seen as a function $\widehat{\mathcal{R}}_{\mathcal{D}_{\text{train}}}(\theta)$ of the parameters θ. Usually, training algorithms for neural networks, given an initial set of parameters $\theta^{(0)} \in \Theta$, generate a discrete weight trajectory $\left(\theta^{(i)}\right)_{i=1}^{N}$ by minimizing the empirical risk $\widehat{\mathcal{R}}_{\mathcal{D}_{\text{train}}}(\theta)$ iteratively with respect to the parameters θ until some stopping criteria are satisfied. The output of the algorithm is thus the neural network with architecture a and parameters given by one of the weights $\theta^{(i)}$ generated in the trajectory, usually the last one, $\theta^{(N)}$. In this work, we assume that neural networks are trained in this way.

2.3 Convolutional Neural Networks

Convolutional neural networks (CNNs) are a special family of FCFNNs with a specific architecture designed to work with data that have a grid-like structure, such as images. CNNs are one of the most common type of networks used in computer vision. They implement convolutional and pooling transformations, that are special functions between layers that impose restrictions on the set of weights and on the activation functions. A layer l whose output is the result of a convolutional or pooling transformation is called a convolutional or pooling layer, respectively.

A *convolutional layer* is a layer whose activations are the result of performing convolutions between the values associated to the neurons of its input layer rearranged as a grid and a filter tensor that imposes a set of equalities on the coefficients of the weight matrices before applying activation functions. In one of the most typical scenarios, a convolutional transformation is a transformation between the activations of V_{l-1} and V_l in such a way that V_{l-1} and V_l are ordered in grid structures of size $h^{(l-1)} \times w^{(l-1)} \times c^{(l-1)}$ and $h^{(l)} \times w^{(l)} \times c^{(l)}$, respectively. We say that a layer l arranged in this grid structure has *height* $h^{(l)}$, *width* $w^{(l)}$, and $c^{(l)}$ *channels*, where each channel is defined as the subgrid obtained by fixing one value in $\left[c^{(l)}\right]$ for the third dimension.

The output of a convolutional layer before applying its activation function φ is computed as follows. First, for each $j \in [c^{(l)}]$, a total of $c^{(l-1)}$ discrete convolutions are performed for the activations of each channel $k \in [c^{(l-1)}]$ of size $h^{(l-1)} \times w^{(l-1)}$ of the layer $l-1$. For this purpose, $c^{(l-1)}$ convolutional filters $C_{k,j}^{(l)}$ of size $k_h^{(l)} \times k_w^{(l)}$ are used. The values of the convolutional filters are called convolutional *filter weights*, and correspond with the parameters of layer l.

Depending on the application, bias terms can be added to the result of the convolutions. For a fixed j, the $c^{(l-1)}$ convolutions generate $c^{(l-1)}$ grids of size $h^{(l)} \times w^{(l)}$. These grids are summed to obtain the activations of the j-th channel of the layer l, obtaining a final grid of size $h^{(l)} \times w^{(l)} \times c^{(l)}$. Summarizing, a typical convolutional layer calculates the (r, s, j)-component of the l-th grid before applying the activation function as

$$\phi_{\mathcal{N}}^{(l)}(x)_{r,s,j} = b_j^{(l)} + \sum_{k=1}^{c^{(l-1)}} \sum_{u=1}^{k_h^{(l)}} \sum_{v=1}^{k_w^{(l)}} C_{k,j}^{(l)}(u,v,) \, \phi_{\mathcal{N}}^{(l-1)}(x)_{r+u-1,\,s+v-1,\,k}$$

where the terms $b_j^{(l)}$ are the bias terms of the layer l for each channel $j \in [c^{(l)}]$.

Pooling transformations are similar operations aiming to reduce the dimensionality of their input layer. One of the most common pooling transformations is the *max pooling* transformation. Given the input layer of the pooling transformation structured as a grid of size $h^{(l-1)} \times w^{(l-1)} \times c^{(l-1)}$, the output of the max pooling transformation is a grid of size $h^{(l)} \times w^{(l)} \times c^{(l)}$ where $c^{(l-1)} = c^{(l)}$ and where each value for a fixed channel $j \in [c^{(l)}]$ is the maximum of a subgrid of size $k_h^{(l)} \times k_w^{(l)}$ for the same channel j of the input grid. The subgrid associated to each (r, s, j)-component of the output grid is chosen using a sliding window of size $k_h^{(l)} \times k_w^{(l)}$ that moves either $s_h^{(l)}$ positions vertically or $s_w^{(l)}$ positions horizontally at each step in such a way the full input grid is covered. A typical max pooling transformation calculates the (r, s, j)-component of the l-th grid as

$$\phi_{\mathcal{N}}^{(l)}(x)_{r,s,j} = \max_{u \in [k_h^{(l)}]} \max_{v \in [k_w^{(l)}]} \phi_{\mathcal{N}}^{(l-1)}(x)_{(r-1)s_h^{(l)}+u,\,(s-1)s_w^{(l)}+v,\,j}.$$

In general, not all components of input values are used to compute the activations of a specific neuron in a convolutional or pooling layer. The region of the input, or more specifically, the set of input vector components, that affect the activations of a neuron is called its *receptive field*. A thorough introduction to CNNs can be found in Goodfellow et al. (2016, Chapter 9).

2.4 Decision Regions

Fully connected feedforward networks define functions $\phi_{\mathcal{N}} \colon \mathbb{R}^{N_0} \to \mathbb{R}^{N_L}$, which is suitable whenever $\mathcal{X}$ and $\mathcal{Y}$ are subsets of Euclidean spaces. However, for classification tasks, one has $\mathcal{Y} = [k]$ for some $k \in \mathbb{N}$. In this case, one selects a projection function $\pi \colon \mathbb{R}^{N_L} \to [k]$ that projects the outputs of the neural network $\mathcal{N}$ into the finite set of points $[k]$ and one uses $\pi \circ \phi_{\mathcal{N}} \colon \mathbb{R}^{N_0} \to [k]$ as a model. In this context, the *decision region* of a label $y \in [k]$ is the set of inputs x such that $\pi(\phi_{\mathcal{N}}(x)) = y$.

To train the neural network, it is common to choose surrogate loss functions $\mathcal{L}^{\mathrm{surr}}$ that depend directly on the output of $\phi_{\mathcal{N}}$. In classification tasks, the most common configuration for a problem with $\mathcal{X}$ having $\mathbb{R}^d$ as an ambient space and $\mathcal{Y} = [k]$ is

Fig. 2.2 Decision regions and boundaries for three different classification problems ordered from left to right by increasing complexity, with three labels each. Black lines represent decision boundaries given by FCFNNs. Decision regions and their boundaries give valuable information on the neural network used in each case. For the two first problems, the decision regions and boundaries are "simple", in the sense that they seem to properly classify the inputs of the domain without visible outliers or "strange" regions. In the second problem, the blue decision region has one hole that does not exist in the first problem. In the third problem, the blue decision region has two connected components, and the red and green regions have protuberances that could have appeared due to outliers in the training data. The extra connected component is detected by the homology of the blue decision region; see Chap. 3 for details

to choose neural networks with $N_0 = d$, $N_L = k$, projection

$$\pi(x_1, \ldots, x_k) = \mathrm{argmax}_{i \in [k]} \, x_i$$

and surrogate loss function $\mathcal{L}^{\mathrm{surr}}$ depending only on ϕ_N, like the *categorical cross entropy loss* CE, that, in its most basic form, looks like

$$\mathrm{CE}(\phi_N, x, y) = -\frac{1}{k} \log \left(\frac{\exp(\phi_N(x)_y)}{\sum_{i=1}^{k} \exp(\phi_N(x)_i)} \right).$$

For neural networks with $N_L = k$ and $\pi(x_1, \ldots, x_k) = \mathrm{argmax}_{i \in [k]} \, x_i$, there is an ambiguity when the argmax is not unique, i.e., when there are at least two indices $i, j \in [k]$ such that $x_i = x_j$. In this case, the projection function must define a deterministic way to choose one of the equal-valued indices. Given a neural network N as the previous one, the set of points $x \in \mathcal{X}$ for which there exists an ambiguity constitutes the *decision boundary* of N. Formally,

$$\mathcal{B}(N) = \left\{ x \in \mathrm{Dom}(\phi_N) : \exists\, i, j \in [N_L] \text{ such that} \right.$$

$$\left. \phi_N(x)_i = \phi_N(x)_j = \max_{y \in [k]} \phi_N(x)_y \right\}. \tag{2.4}$$

Sometimes, decision regions are studied without their decision boundary; in other words, $(\pi \circ \phi_N)^{-1}(y) \setminus \mathcal{B}(N)$ is used. Figure 2.2 shows examples of decision boundaries and regions for classification problems with three labels.

In this context, it is usual to interpret the outputs of the function ϕ_N as (unnormalized) probabilities, indicating the likelihood that the input belongs to one of the classes represented by the outputs. Specifically, $\phi_N(x)_i$ denotes the (unnormalized) probability that x belongs to the class $i \in [k]$. Therefore, π is chosen as argmax since it selects the class with the highest probability according to the output of the neural network.

For binary classification problems, i.e., $\mathcal{Y} = [2]$, the usual neural network architectures, projection functions, and decision boundaries are slightly different. On the one hand, the number of neurons in the last layer is set to one, that is, $N_L = 1$. On the other hand, the usual projection function is given by $\pi(x) = 2$ if $x > b$ and $\pi(x) = 1$ if $x \leq b$, with the usual value of b being zero. In this configuration, the decision boundary is defined as

$$\mathcal{B}(\mathcal{N}) = \{x \in \mathrm{Dom}(\phi_\mathcal{N}) : \phi_\mathcal{N}(x) = b\}. \tag{2.5}$$

References

Michael M. Bronstein, Joan Bruna, Taco Cohen, and Petar Veličković. Geometric deep learning: Grids, groups, graphs, geodesics, and gauges, 2021. URL https://arxiv.org/abs/2104.13478.

Vincent François-Lavet, Peter Henderson, Riashat Islam, Marc G. Bellemare, and Joelle Pineau. An introduction to deep reinforcement learning. *Foundations and Trends®in Machine Learning*, 11(3–4):219–354, 2018. ISSN 1935-8237. URL http://dx.doi.org/10.1561/2200000071.

Ian Goodfellow, Yoshua Bengio, and Aaron Courville. *Deep Learning*. MIT Press, 2016. http://www.deeplearningbook.org.

Philipp Grohs and Gitta Kutyniok. *Mathematical Aspects of Deep Learning*. Cambridge University Press, 2022. https://doi.org/10.1017/9781009025096.

Charline Le Lan. *Understanding representation learning for deep reinforcement learning*. PhD thesis, University of Oxford, 2023.

Volodymyr Mnih, Koray Kavukcuoglu, David Silver, Andrei A. Rusu, Joel Veness, Marc G. Bellemare, Alex Graves, Martin Riedmiller, Andreas K. Fidjeland, Georg Ostrovski, Stig Petersen, Charles Beattie, Amir Sadik, Ioannis Antonoglou, Helen King, Dharshan Kumaran, Daan Wierstra, Shane Legg, and Demis Hassabis. Human-level control through deep reinforcement learning. *Nature*, 518(7540):529–533, Feb 2015. ISSN 1476-4687. URL https://doi.org/10.1038/nature14236.

Simon J. D. Prince. *Understanding Deep Learning*. MIT Press, 2023. URL http://udlbook.com.

Chapter 3
Topological Data Analysis

Abstract A variety of tools from topological data analysis have been used within machine learning. Among them, persistent homology and Mapper graphs have so far been the most relevant in the study of deep learning methods. In this chapter, we provide a concise overview of both.

3.1 Persistent Homology

The book by Munkres (2000) is a standard reference on general topology. For an introduction to computational topology, we refer the reader to Edelsbrunner and Harer (2022).

An *abstract simplicial complex* is a collection K of non-empty finite subsets of a set P such that $\{p\}$ is in K for all $p \in P$ and such that σ is in K whenever $\sigma \subseteq \sigma'$ with σ' in K. Elements of P are called *vertices* and elements of K are called *simplices*. A simplex σ has *dimension d* if its cardinality is $d + 1$, and the dimension of a finite simplicial complex K is the maximum dimension of its simplices. From now on we omit the term "abstract", since geometric simplicial complexes will not be discussed in this book.

Every undirected simple graph $G = (V, E)$ with set of vertices V and set of edges E can be viewed as a 1-dimensional simplicial complex with $P = V$ and a simplex $\sigma_e = \{v, w\}$ for every edge $e \in E$ joining v and w.

3.1.1 Simplicial Homology

Each finite simplicial complex K with a prescribed order on its set of vertices has an associated family of *homology groups* $H_n(K)$ for $n \in \mathbb{N}$, defined as follows. An *n-chain* is a finite sum $c = \sum_i a_i \sigma_i$ of n-dimensional simplices σ_i of K with coefficients a_i in the ring of integers $\mathbb{Z}$.

© The Author(s), under exclusive license to Springer Nature Switzerland AG 2026
R. Ballester et al., *Topological Data Analysis for Neural Networks*, SpringerBriefs in Computer Science, https://doi.org/10.1007/978-3-032-08283-1_3

The *boundary* of an ordered n-simplex $\sigma = (v_0, \ldots, v_n)$ is the alternating sum

$$\partial \sigma = \sum_{i=0}^{n} (-1)^i \, (v_0, \ldots, \widehat{v_i}, \ldots, v_n),$$

where $\widehat{v_i}$ means that v_i is omitted. The boundary operator is linearly extended to n-chains, that is, if $c = \sum_i a_i \sigma_i$ then $\partial c = \sum_i a_i \partial \sigma_i$.

An n-chain c is an n-*cycle* if $\partial c = 0$. The n-*th homology group* of K, denoted $H_n(K)$, is defined as a quotient of the abelian group of n-cycles modulo the subgroup of n-boundaries, that is, n-chains of the form ∂c for some $(n+1)$-chain c. These are indeed n-cycles, since $\partial \partial c = 0$ for all c.

If coefficients in a field $\mathbb{F}$ are used to define n-chains, then $H_n(K)$ becomes an $\mathbb{F}$-vector space. For convenience, we will work with such vector spaces in most of what follows. If no coefficient field is explicitly selected, we will assume that the one used is the field $\mathbb{F}_2$ of two elements.

The dimension of $H_n(K)$ as an $\mathbb{F}$-vector space is called the n-*th Betti number* of K and denoted by $\beta_n(K)$. The zeroth Betti number $\beta_0(K)$ counts the number of connected components of K. For $n \geq 1$, generators of $H_n(K)$ are interpreted as n-dimensional "cavities" in the geometric realization of K. For example, if K is a triangulation of a circle, then $\beta_1(K) = 1$, while if K is a triangulation of a 2-sphere, then $\beta_1(K) = 0$ and $\beta_2(K) = 1$, and if K is a triangulation of a torus, then $\beta_1(K) = 2$ and $\beta_2(K) = 1$. For a more detailed discussion of simplicial homology, see Edelsbrunner and Harer (2022).

3.1.2 *Persistence Modules*

An $\mathbb{R}$-indexed *persistence module* is a family $\mathbb{V} = (V_t)_{t \in \mathbb{R}}$ of vector spaces over a field $\mathbb{F}$ equipped with $\mathbb{F}$-linear maps $f_{s,t} \colon V_s \to V_t$ for $s \leq t$, such that $f_{s,t} \circ f_{r,s} = f_{r,t}$ if $r \leq s \leq t$ and $f_{t,t} = \mathrm{id}$ for all t.

Given a filtration of simplicial complexes $(K_t)_{t \in \mathbb{R}}$ ordered by inclusion, the family $(H_*(K_t))_{t \in \mathbb{R}}$ is a persistence module, where $H_*(K_t)$ denotes the direct sum of $H_n(K_t)$ for all $n \in \mathbb{N}$. The $\mathbb{F}$-linear maps $f_{s,t} \colon H_*(K_s) \to H_*(K_t)$ are induced by the inclusions $K_s \subseteq K_t$ if $s \leq t$.

Persistence modules are discussed in detail in Chazal et al. (2016). We denote the persistence module given by the n-th homology H_n of a filtration of simplicial complexes $(K_t)_{t \in \mathbb{R}}$ and coefficient field $\mathbb{F}$ by $\mathbb{V}_n^{\mathbb{F}}(K)$. We drop the superscript denoting the field when it is clear from the context.

A persistence module $\mathbb{V}$ is of *finite type* if V_t is finite-dimensional for all t and, moreover, there is a finite set $\{t_0, \ldots, t_k\} \subset \mathbb{R}$ such that $V_t = 0$ if $t < t_0$ and $f_{s,t}$ is an isomorphism whenever the half-open interval $(s, t]$ does not contain any of the points $t_0, \ldots, t_k$.

A vector $v \in V_b$ is said to be *born* at a parameter value b if it is not in the image of $f_{s,b}$ for any $s < b$, and a vector $v \in V_s$ *dies* at a parameter value $d > s$ if $f_{s,d}(v) = 0$ while $f_{s,t}(v) \neq 0$ for $s \leq t < d$. As explained in Chazal et al. (2016, § 1.5), the lifetime intervals $[b, d) \subset \mathbb{R}$ of a suitable collection of basis elements of a persistence module $\mathbb{V}$ of finite type represent $\mathbb{V}$ up to isomorphism. More precisely, $\mathbb{V}$ is isomorphic to a finite direct sum of *interval modules* $\mathbb{I}[b, d)$, where $I[b, d)_t = \mathbb{F}$ if $b \leq t < d$, and zero otherwise.

The intervals $[b, d)$ form a *multiset*, since they can be repeated, so every interval is given with a multiplicity. This multiset is called a *barcode* of $\mathbb{V}$. We denote by $\bar{\mathbb{R}}$ the extended real line with a point at infinity, and assume that $d \in \bar{\mathbb{R}}$ since death values are infinite in the case of vectors $v \in V_s$ for which $f_{s,t}(v) \neq 0$ for all t.

3.1.3 Persistence Diagrams

Barcodes are more efficiently represented by means of *persistence diagrams*, which are multisets of points in a coordinate plane with a point (b, d) with $b < d$ and $d \in \bar{\mathbb{R}}$ for each interval $[b, d)$ in the barcode of a persistence module $\mathbb{V}$. We denote by $D(\mathbb{V}) = \{(b_i, d_i)\}_{i \in I}$ the persistence diagram of a persistence module $\mathbb{V}$, where I is the multiset of intervals $[b_i, d_i)$ in its associated barcode; see Fig. 3.1.

In the case of a filtration $(K_t)_{t \in \mathbb{R}}$ of simplicial complexes, the persistence diagram of $(H_*(K_t))_{t \in \mathbb{R}}$ describes the evolution of homology generators along the filtration. Thus, a representative n-cycle ζ can be associated with each point (b, d) in homological degree n, so that b is the birth parameter of ζ and d is the value at which ζ becomes a boundary.

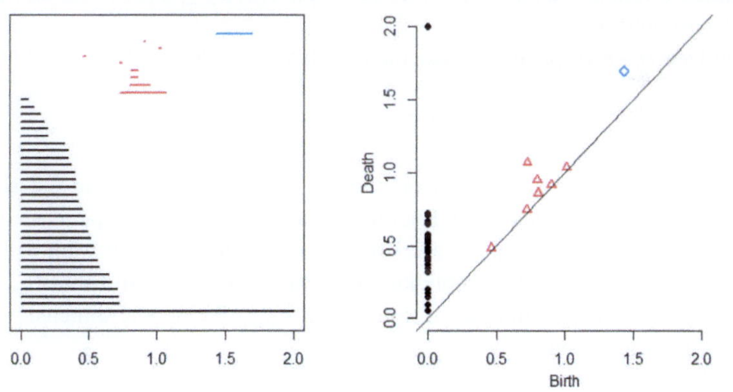

Fig. 3.1 A barcode and its associated persistence diagram: each interval $[b, d)$ in the barcode is represented as a point (b, d) in a coordinate system. The barcode corresponds to a Vietoris–Rips filtration (Sect. 3.1.4) of a point cloud with 30 points sampled from the surface of a 3D sphere of radius 1. Black: H_0; red: H_1; blue: H_2

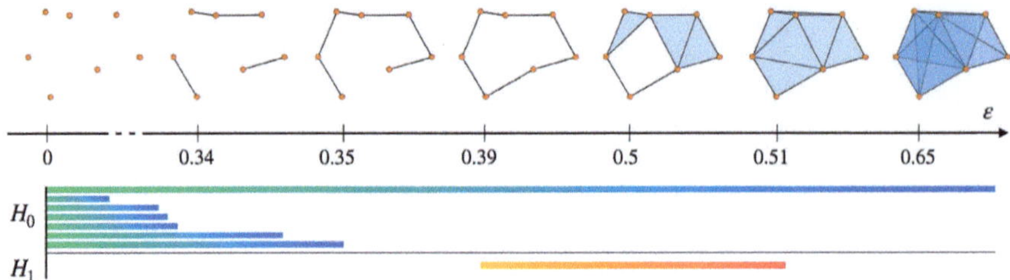

Fig. 3.2 Barcode from a Vietoris–Rips filtration of a point cloud in the plane. The image is part of Figure 1 in Guerra et al. (2021)

We briefly mention an extension of the usual persistence modules computed with homology that also produce persistence diagrams, known as *zigzag persistence*, introduced by Carlsson and de Silva (2010). Zigzag persistence modules are analogous to persistence modules, but the inclusions are not necessarily given by the order of the real numbers indexing the vector spaces. Zigzag persistence modules share a good amount of properties with ordinary persistence modules, such as their unique representation by persistence diagrams or their stability (Cohen-Steiner et al. 2007; Botnan and Lesnick 2018), under mild assumptions.

3.1.4 Filtrations

There are several ways to construct filtrations of simplicial complexes given a point cloud or a weighted graph. In this book, a *point cloud* is a finite set P equipped with a symmetric function $d \colon P \times P \to \bar{\mathbb{R}}$ such that $d(x, x) \leq d(x, y)$ for all $x, y \in P$, called a *dissimilarity*.

The *Vietoris–Rips filtration* associated with a point cloud (P, d) is defined as

$$\mathrm{VR}_t(P, d) = \{\sigma \subseteq P \, : \, \sigma \neq \emptyset, \, \mathrm{diam}(\sigma) \leq t\},$$

where $\mathrm{diam}(\sigma)$ is the supremum of $d(x, y)$ for $x, y \in \sigma$. Figure 3.2 shows an example with seven points in the Euclidean plane, with the corresponding barcode.

For practical purposes, there are situations in which we may want to limit the dimension of Vietoris–Rips simplicial complexes to a maximum value $k_{\max}$. In this case, we only take simplices up to dimension $k_{\max}$. We denote dimension-limited Vietoris–Rips filtrations by

$$\mathrm{VR}_t^{k_{\max}}(P, d) = \{\sigma \in \mathrm{VR}_t(P, d) \, : \, \dim(\sigma) \leq k_{\max}\}.$$

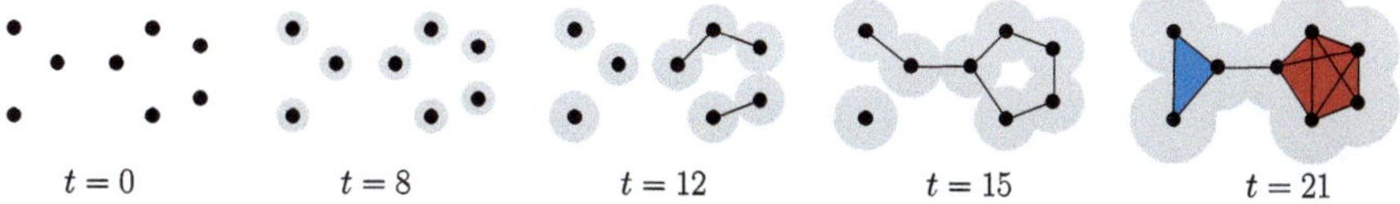

$$t = 0 \qquad t = 8 \qquad t = 12 \qquad t = 15 \qquad t = 21$$

Fig. 3.3 Čech filtration at time values $t \in \{0, 8, 12, 15, 21\}$ for a point cloud in the plane. For $t \leq 15$, only edges are added as there are no mutually intersecting three circles. When $t = 21$, one triangle and two tetrahedra have emerged in the filtration

If the point cloud P is a subset of a metric space (X, d), then another popular family of simplicial complexes is the *Čech filtration*

$$\check{C}_t(P, X, d) = \left\{ \sigma \subseteq P \, : \, \sigma \neq \emptyset, \, \bigcap_{x \in \sigma} \bar{B}(x, t/2) \neq \emptyset \right\},$$

for $t \geq 0$, where $\bar{B}(x, \varepsilon) = \{y \in X : d(x, y) \leq \varepsilon\}$ denotes the closed ball centered at x of radius ε. We assume that $\check{C}_t(P, X, d) = \emptyset$ if $t < 0$. The indexing parameter has been chosen so that $\check{C}_t(P, X, d) \subseteq \mathrm{VR}_t(P, d)$ for all t. In fact, $\check{C}_t(P, X, d)$ and $\mathrm{VR}_t(P, d)$ have the same edges for all t. Figure 3.3 shows an example in the Euclidean plane.

3.1.5 Weighted Graphs

Similar ideas can be applied to weighted graphs. Let (G, w_V, w_E) be a weighted graph, where $w_V \colon V(G) \to \mathbb{R}$ and $w_E \colon E(G) \to \mathbb{R}$ are weight functions for the vertices and the edges of G, respectively. By requiring that $w_V(v) \leq w_E(e)$ for all $v \in V(G)$ and all $e \in E(G)$ incident to v, the weighted graph G can be treated as a point cloud. Points correspond to the vertices of G and distances between points are given by the weight functions w_V and w_E. The lack of an edge between two vertices is encoded as an infinite distance. Thus the distance from a point to itself need not be zero:

$$d(v, w) = \begin{cases} w_V(v) & \text{if } v = w, \\ w_E(\{v, w\}) & \text{if } \{v, w\} \in E(G), \\ \infty & \text{otherwise.} \end{cases} \tag{3.1}$$

This allows one to compute a Vietoris–Rips filtration in the same way as for point clouds. We denote the Vietoris–Rips filtration of a weighted graph (G, w_V, w_E) by $\mathrm{VR}(G, w_V, w_E)$, and we let $\mathrm{VR}^{k_{\max}}(G, w_V, w_E)$ be the dimension-limited version.

Sometimes, weighted graphs are provided only with weights on their edges. In such cases, there is no canonical way to define weights on vertices. One solution is to define w_V for a vertex $v \in V(G)$ as the minimum weight of all its incident edges, and another possible solution is to define w_V as the global minimum weight among all edge weights.

Vietoris–Rips and Čech filtrations incorporate simplices with lower diameter first, that is, the lower the values of t, the lower the diameter of the simplices inside VR_t and $\check{C}_t$. However, it is usual in the graph realm to add edges with higher weights first. Thus, by giving a descending order $e_1, \ldots, e_n$ of the edges by their weights, where $n = |E(G)|$, the edge e_i is added at $t = i - 1$ to the filtration for all $i \in [n]$. In this case, the vertices can be included either at $t = 0$, or at the moment the first incident edge enters the filtration, or at $t = w_V(v)$ if the vertices are weighted by a function w_V that satisfies $w_V(v) \geq w_E(e)$ for all vertices v and all edges e incident to v. In the first two cases, we define the following dissimilarity functions:

$$d_{\downarrow}^0(v, w) = \begin{cases} 0 & \text{if } v = w, \\ i - 1 & \text{if } \{v, w\} = e_i, \\ \infty & \text{otherwise}, \end{cases}$$

$$d_{\downarrow}(v, w) = \begin{cases} \min\{i : v \in e_i\} & \text{if } v = w, \\ i - 1 & \text{if } \{v, w\} = e_i, \\ \infty & \text{otherwise}. \end{cases}$$

In the third case, by defining $\mathfrak{S} = E(G) \cup \{\{v\} : v \in V(G)\}$ and $w \colon \mathfrak{S} \to \mathbb{R}$ as

$$w(\{v, w\}) = \begin{cases} w_V(v) & \text{if } v = w, \\ w_E(\{v, w\}) & \text{otherwise}, \end{cases}$$

and considering the elements $s_1, \ldots, s_m$ of $\mathfrak{S}$ in descending order as given by the values of w, we use the dissimilarity function

$$d_{\downarrow}^V(v, w) = \begin{cases} i - 1 & \text{if } \{v, w\} = s_i, \\ \infty & \text{if } \{v, w\} \notin \mathfrak{S}. \end{cases}$$

For points (b, d) in persistence diagrams associated with Vietoris–Rips filtrations, b and d are values of the corresponding dissimilarity function for some pair of points in the point cloud. This means that points of Vietoris–Rips persistence diagrams coming from functions $d_{\downarrow}^0$, $d_{\downarrow}$, and $d_{\downarrow}^V$ are tuples of indices of edges in the graph. Therefore, an alternative persistence diagram $D^w(\mathbb{V}_k(\mathrm{VR}(\cdot)))$ containing the weights of the edges can be obtained by taking the weight values of the edges associated with the indices in the persistence diagram. That is,

$$D^w(\mathbb{V}_k(\mathrm{VR}(V(G), d)))$$

$$= \left\{ (w_E(e_{i+1}), w_E(e_{j+1})) : (i, j) \in D(\mathbb{V}_k(\mathrm{VR}(V(G), d))) \right\} \qquad (3.2)$$

for $d \in \{d_\downarrow^0, d_\downarrow\}$, and

$$D^w\big(\mathbb{V}_k\big(\mathrm{VR}\big(V(G), d_\downarrow^V\big)\big)\big)$$

$$= \big\{(w(s_{i+1}), w(s_{j+1})) : (i, j) \in D(\mathbb{V}_k(\mathrm{VR}(V(G), d)))\big\}, \tag{3.3}$$

where we define $w_E(e_\infty) = w(s_\infty) = \infty$. We refer to this persistence diagram as the *weighted persistence diagram* of the persistence modules of the Vietoris–Rips family of filtrations for graphs induced by $d_\downarrow^0$, $d_\downarrow$ or $d_\downarrow^V$. This also holds for dimension-limited Vietoris–Rips filtrations.

There are many useful filtrations that can be employed in persistent homology for finite sets of points or graphs. In this section, we have presented only some of them. For a broader perspective on point cloud and graph filtrations, we refer the reader to the works by Chazal et al. (2014), Edelsbrunner and Harer (2022), and Ballester and Rieck (2024).

3.1.6 Morse Functions

Compact manifolds also yield persistence diagrams computed from Morse functions. A *Morse function* is a smooth function $f : M \to \mathbb{R}$ on a smooth manifold M such that all critical points are non-degenerate, that is, the Hessian of f at each critical point is non-singular. Similarly to the simplicial case, one can build persistence modules from the homology of sublevel sets of a Morse function at different levels $t \in \mathbb{R}$. Thus, given $n \in \mathbb{N}$ and a field $\mathbb{F}$, the persistence module $\mathbb{V}_n(f)$ for a Morse function f on a smooth manifold M is defined as

$$V_n(f)_t = H_n(f^{-1}(-\infty, t]),$$

where singular homology with coefficients in $\mathbb{F}$ is meant. If M is compact, then $\mathbb{V}_n(f)$ is of finite type and hence it has a barcode and a persistence diagram.

3.1.7 Wasserstein Distances

For many purposes, it is useful to have a notion of distance between persistence diagrams. The most frequently used distances between persistence diagrams are the bottleneck distance and the Wasserstein distances.

The *bottleneck distance* between two persistence diagrams D_1 and D_2 is

$$W_\infty(D_1, D_2) = \inf_{\eta : D_1^\Delta \to D_2^\Delta} \sup_{x \in D_1^\Delta} \| x - \eta(x) \|_\infty,$$

where D_1^Δ and D_2^Δ denote the persistence diagrams D_1 and D_2 supplemented with their diagonals, that is, $D_1^\Delta = D_1 \cup \Delta$ and $D_2^\Delta = D_2 \cup \Delta$, where $\Delta = \{(x, x) : x \in \bar{\mathbb{R}}\}$, and $\eta \colon D_1^\Delta \to D_2^\Delta$ ranges over all possible bijections of multisets, where points in Δ have countably infinite multiplicity. The norm $\| \cdot \|_\infty$ is the supremum norm.

The q-th *Wasserstein distance* W_q is defined as

$$W_q(D_1, D_2) = \inf_{\eta \colon D_1^\Delta \to D_2^\Delta} \left(\sum_{x \in D_1^\Delta} \|x - \eta(x)\|_\infty^q \right)^{1/q},$$

although a p-norm $\| \cdot \|_p$ may be used instead of the supremum norm $\| \cdot \|_\infty$, in which case we denote the corresponding Wasserstein distance by $W_q^p(D_1, D_2)$.

3.1.8 Persistence Summaries

Persistence diagrams are not used directly for data analysis due to their lack of a suitable structure for statistical inference. Instead, persistence summaries are used, which are real-valued functions or vector-valued functions derived from persistence diagrams. The following are examples of numerical persistence summaries.

The *total persistence* of a persistence diagram $D = \{(b_i, d_i)\}_{i \in I} \cup \{(b_j, \infty)\}_{j \in J}$ is the sum of the lifetimes of the non-infinity points:

$$\mathrm{TP}(D) = \sum_{i \in I}(d_i - b_i),$$

and, for $p \geq 1$, the *p-norm* is a generalization of total persistence given by

$$\|D\|_p = \left(\sum_{i \in I}(d_i - b_i)^p \right)^{1/p}.$$

The most frequently used vector-valued summaries are Betti curves, persistence landscapes, and persistence images, which are described next.

Given a persistence diagram $D = \{(b_i, d_i)\}_{i \in I} \cup \{(b_j, \infty)\}_{j \in J}$, the *Betti curve* of D (Fig. 3.4) is the graph of the function $\beta \colon \mathbb{R} \to \mathbb{N}$ defined as

$$\beta(t) = \left| \{i \in I : b_i \leq t \leq d_i\} \right|.$$

If the diagram D comes from persistent n-th homology H_n of a filtered simplicial complex $(K_t)_{t \in \mathbb{R}}$ with coefficients in a field $\mathbb{F}$, then $\beta(t)$ counts how many homology generators remain alive at parameter value t. In other words, $\beta(t) = \beta_n(K_t)$.

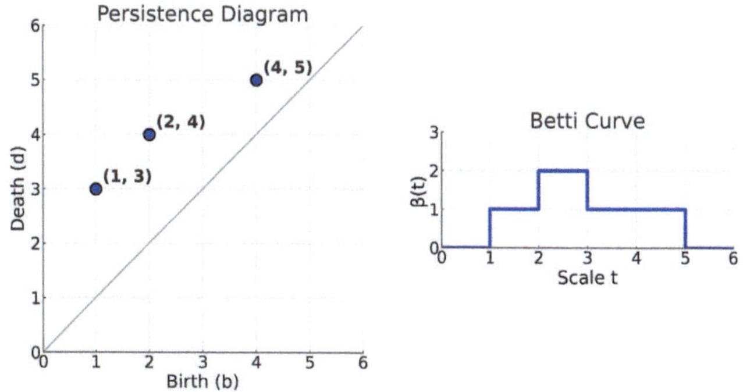

Fig. 3.4 Betti curve (right) drawn from a persistence diagram (left)

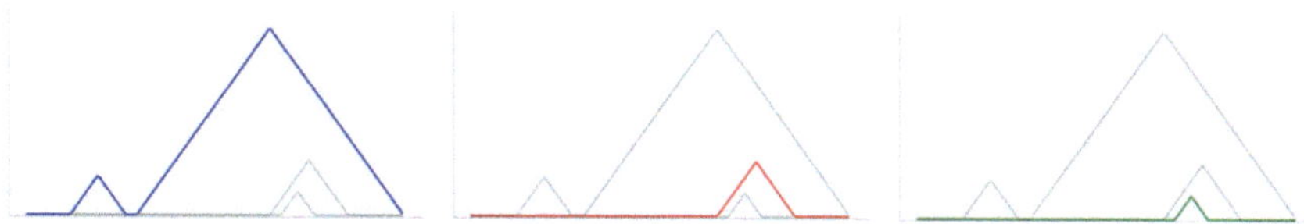

Fig. 3.5 Three consecutive levels λ_1, λ_2 and λ_3 of a persistence landscape

A *landscape* (Bubenik 2015) of a persistence diagram D is a sequence of piecewise linear functions $\lambda_k \colon \mathbb{R} \to \mathbb{R}$ for $k \geq 1$, defined as

$$\lambda_k(t) = \max_{i \in I}^{k} \left\{ f_{(b_i, d_i)}(t) \right\},$$

where $\max^k$ returns the k-th maximum value of a multiset, and

$$f_{(b,d)}(t) = \max(0, \min(b + t, d - t))$$

is a *tent* function with apex $\left(\frac{1}{2}(b + d), \frac{1}{2}(d - b) \right)$, as depicted in Fig. 3.5.

Persistence images were defined in Adams et al. (2017). A surface is generated by centering 2-dimensional Gaussian distributions at each non-infinity point of a given persistence diagram $D = \{(b_i, d_i)\}_{i \in I} \cup \{(b_j, \infty)\}_{j \in J}$, as follows:

$$\rho_D(x, y) = \sum_{i \in I} f(b_i, d_i - b_i)\, \phi_i(x, y)$$

for $(x, y) \in \mathbb{R}^2$, where f is a continuous nonnegative weighting function vanishing along the horizontal axis, and ϕ_i is a normal distribution centered at $(b_i, d_i - b_i)$, for each $i \in I$, with a fixed variance. Then, a *persistence image* is obtained by integrating ρ_D over each pixel (Fig. 3.6).

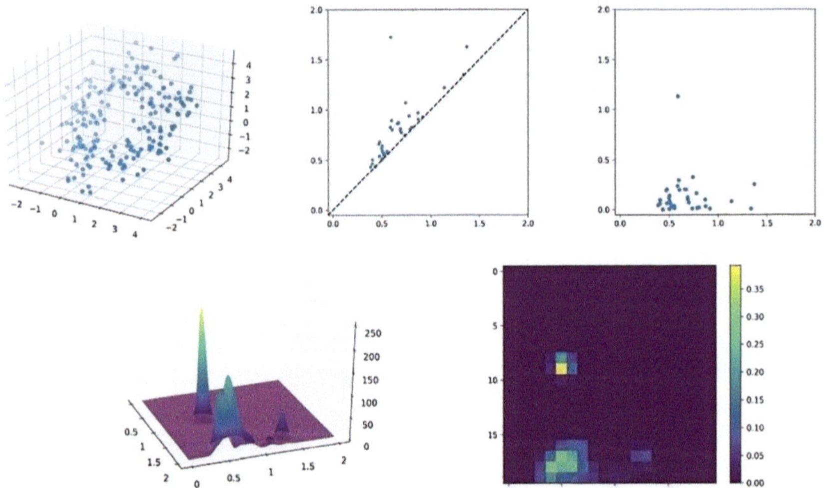

Fig. 3.6 Persistence image from a point cloud. The transformation $T(x, y) = (x, y - x)$ is applied to the persistence diagram, sending the diagonal to the horizontal axis, before picking a Gaussian distribution with fixed variance centered at each point. The images are part of Figure 7 in Stolz et al. (2021)

3.2   Mapper

Mapper is an algorithm, introduced by Singh et al. (2007), that extracts visual descriptions of high-dimensional datasets in the form of simplicial complexes (usually graphs). One key concept behind Mapper is the nerve of a covering. Given a topological space X and a covering $\mathcal{U} = \{\mathcal{U}_i\}_{i \in I}$ of X, the *nerve* of $\mathcal{U}$ is defined as the simplicial complex $N(\mathcal{U})$ whose set of vertices is I and where a family $\{i_1, \ldots, i_k\} \subseteq I$ is a simplex of $N(\mathcal{U})$ if and only if $\bigcap_{r=1}^{k} \mathcal{U}_{i_r} \neq \emptyset$. For *good* coverings (i.e., open coverings whose sets and their non-empty finite intersections are contractible), the geometric realization of $N(\mathcal{U})$ is homotopy equivalent to X. Hence, $N(\mathcal{U})$ encodes relevant features of the shape of X into a combinatorial object.

Mapper applies the nerve construction to suitably customized coverings of datasets. Simplicial complexes produced by Mapper are often able to reveal patterns of a latent topological space from which the dataset might have been sampled. The Mapper algorithm has the following ingredients as input:

1. A dataset $\mathcal{D}$, which we assume of finite cardinality;
2. A filter function $f : \mathcal{D} \rightarrow \mathbb{R}^d$, where $d = 1$ or $d = 2$ in most use cases;
3. A finite cover $\mathcal{U} = \{\mathcal{U}_i\}_{i \in I}$ of $\mathrm{Im}(f) \subset \mathbb{R}^d$ consisting of overlapping open sets;
4. A clustering algorithm, such as DBSCAN (Ester et al. 1996).

The method is described in Algorithm 1 in the Appendix.

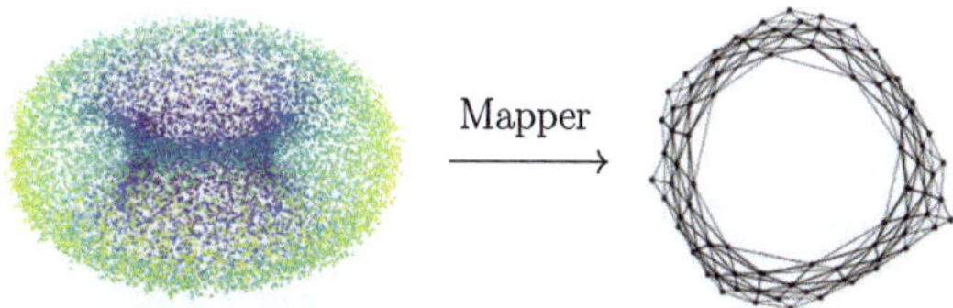

Fig. 3.7 Mapper graph generated from a noisy sample of a torus with 14,400 points using Kepler Mapper (van Veen et al. 2019). The filter function used is given by the projections to the x and y axes. Covers are taken with an overlap of $p = 0.2$ using 100 squares, by means of the standard cover implementation of the software

Mapper is most often applied using filter functions $f \colon \mathcal{D} \to \mathbb{R}$ and choosing covers $\mathcal{U}$ consisting of intervals with a fixed percentage of overlap p. With these choices and using clustering algorithms that generate disjoint clusters, the simplicial complexes produced by Mapper are graphs (Fig. 3.7). Mapper graphs are closely related with Reeb graphs (Reeb 1946; Edelsbrunner and Harer 2022, Ch. VI). For an advanced analysis of the Mapper algorithm for filter functions $f \colon \mathcal{D} \to \mathbb{R}$ and its connection with Reeb graphs, we refer the reader to the work by Carrière and Oudot (2018).

It was observed by Liu et al. (2023) that, in some experiments studying the output space of neural networks, Mapper produced too many tiny components that did not carry useful information for the experiments. For this reason, they introduced a new algorithm called *graph-based topological data analysis* (GTDA). The GTDA algorithm builds on the Mapper algorithm to construct Reeb networks from graph inputs instead of point clouds. Reeb networks generalize Reeb graphs. Each vertex in a Reeb network built using GTDA is associated with a subgraph of the original graph. Given an input graph G, GTDA uses filter functions $f \colon V(G) \to \mathbb{R}^d$ to build Reeb network vertices using a recursive splitting strategy based on filter function values. After initial vertices are generated, the smallest ones are merged into other vertices and edges are added. Vertices within connected components with non-empty intersection are connected, and other edges are added to promote connectivity of the output. The GTDA algorithm is described in Liu et al. (2023, Algorithm 1).

References

Henry Adams, Tegan Emerson, Michael Kirby, Rachel Neville, Chris Peterson, Patrick Shipman, Sofya Chepushtanova, Eric Hanson, Francis Motta, and Lori Ziegelmeier. Persistence images: A stable vector representation of persistent homology. *Journal of Machine Learning Research*, 18(8):1–35, 2017. URL http://jmlr.org/papers/v18/16-337.html.

Rubén Ballester and Bastian Rieck. On the expressivity of persistent homology in graph learning. In *Third Learning on Graphs Conference*, 2024. URL https://arxiv.org/abs/2302.09826.

Magnus Botnan and Michael Lesnick. Algebraic stability of zigzag persistence modules. *Algebraic & Geometric Topology*, 18(6):3133–3204, oct 2018. URL https://doi.org/10.2140%2Fagt.2018.18.3133.

Peter Bubenik. Statistical topological data analysis using persistence landscapes. *Journal of Machine Learning Research*, 16(3):77–102, 2015. URL http://jmlr.org/papers/v16/bubenik15a. html.

Gunnar Carlsson and Vin de Silva. Zigzag persistence. *Foundations of Computational Mathematics*, 10(4):367–405, Aug 2010. ISSN 1615-3383. URL https://doi.org/10.1007/s10208-010-9066-0.

Mathieu Carrière and Steve Oudot. Structure and stability of the one-dimensional Mapper. *Foundations of Computational Mathematics*, 18(6):1333–1396, Dec 2018. ISSN 1615-3383. URL https://doi.org/10.1007/s10208-017-9370-z.

Frédéric Chazal, Vin de Silva, and Steve Oudot. Persistence stability for geometric complexes. *Geometriae Dedicata*, 173(1):193–214, Dec 2014. ISSN 1572-9168. URL https://doi.org/10. 1007/s10711-013-9937-z.

Frédéric Chazal, Vin de Silva, Marc Glisse, and Steve Oudot. *The Structure and Stability of Persistence Modules*, volume 10 of *SpringerBriefs in Mathematics*. Springer, 2016.

David Cohen-Steiner, Herbert Edelsbrunner, and John Harer. Stability of persistence diagrams. *Discrete & Computational Geometry*, 37(1):103–120, Jan 2007. ISSN 1432-0444. URL https:// doi.org/10.1007/s00454-006-1276-5.

Herbert Edelsbrunner and John L. Harer. *Computational Topology: An Introduction*. American Mathematical Society, 2022.

Martin Ester, Hans-Peter Kriegel, Jörg Sander, and Xiaowei Xu. A density-based algorithm for discovering clusters in large spatial databases with noise. In *Proceedings of the Second International Conference on Knowledge Discovery and Data Mining*, KDD'96, page 226–231. AAAI Press, 1996.

Marco Guerra, Alessandro De Gregorio, Ulderico Fugacci, Giovanni Petri, and Francesco Vaccarino. Homological scaffold via minimal homology bases. *Scientific Reports*, 11(1):5355, 2021. ISSN 2045-2322. URL https://doi.org/10.1038/s41598-021-84486-1.

Meng Liu, Tamal K. Dey, and David F. Gleich. Topological structure of complex predictions. *Nature Machine Intelligence*, Nov 2023. ISSN 2522-5839. URL https://doi.org/10.1038/ s42256-023-00749-8.

James R. Munkres. *Topology*. Prentice Hall, Inc., Upper Saddle River, NJ, 2nd edition, 2000. ISBN 0131816292; 9780131816299; 0131784498; 9780131784499.

Georges Reeb. Sur les points singuliers d'une forme de Pfaff complètement intégrable ou d'une fonction numérique [On the singular points of a completely integrable Pfaff form or of a numerical function]. *Comptes Rendus Acad. Sciences Paris*, 222:847–849, 1946.

Gurjeet Singh, Facundo Mémoli, and Gunnar Carlsson. Topological methods for the analysis of high dimensional data sets and 3D object recognition. In M. Botsch, R. Pajarola, B. Chen, and M. Zwicker, editors, *Eurographics Symposium on Point-Based Graphics*. The Eurographics Association, 2007. ISBN 978-3-905673-51-7. URL https://research.math.osu. edu/tgda/mapperPBG.pdf.

Bernadette J. Stolz, Tegan Emerson, Satu Nahkuri, Mason A. Porter, and Heather A. Harrington. Topological data analysis of task-based fMRI data from experiments on schizophrenia. *Journal of Physics: Complexity*, 2(3):035006, May 2021. URL https://dx.doi.org/10.1088/2632-072X/ abb4c6.

Hendrik Jacob van Veen, Nathaniel Saul, David Eargle, and Sam W. Mangham. Kepler Mapper: A flexible Python implementation of the Mapper algorithm. *Journal of Open Source Software*, 4 (42):1315, 2019. URL https://doi.org/10.21105/joss.01315.

Part II
Interactions

Chapter 4
Challenges in Deep Learning

Abstract Some of the methods for analyzing neural networks discussed in this book have been applied to solve corresponding deep learning problems, which we summarize and contextualize in this chapter.

As mentioned in the Introduction, TDA has been used to analyze various aspects of neural networks, which we have categorized into the following four groups:

1. Architecture;
2. Input and output spaces;
3. Internal representations and activations;
4. Training dynamics and loss functions.

This second part of the book includes a chapter dedicated to each of these four categories. Furthermore, the current chapter describes seven challenges in which TDA has been applied within these categories:

1. Regularization;
2. Pruning;
3. Adversarial, out-of-distribution, and shifted examples;
4. Detection of trojaned networks;
5. Model selection;
6. Prediction of accuracy;
7. Quality assessment of generative models.

Table A.1 in the Appendix contains a list of all the articles included in the survey. For each paper, we specify a one-line summary of it, the categories to which it corresponds, and application directions explored in that paper.

© The Author(s), under exclusive license to Springer Nature Switzerland AG 2026

R. Ballester et al., *Topological Data Analysis for Neural Networks*, SpringerBriefs in Computer Science, https://doi.org/10.1007/978-3-032-08283-1_4

4.1 Regularization

In machine learning, regularization techniques are "algorithmic tweaks" intended to reward models of lower complexity (Zhang et al. 2021) during training. In the context of deep learning, regularization techniques aim to reduce model overfitting, that is, to avoid neural networks that excel in minimizing empirical risk $\widehat{\mathcal{R}}_{\mathcal{D}_{\text{train}}}$ but generalize poorly to a low real risk $\mathcal{R}$.

There are many ways in which regularization is performed in deep learning. Two of the most important ones are *early stopping* and the use of *regularization terms*.

In early stopping, training is stopped whenever a quality measure (usually the empirical risk) evaluated on a set $\mathcal{D}_{\text{val}}$ different from the training dataset $\mathcal{D}_{\text{train}}$, stops improving. Defining the quality measure and the moment at which we consider that this measure has stopped improving is not an easy task, and it depends on the properties of the neural network and the training procedure. Some (non-topological) tips and tricks to tune early stopping can be found in Prechelt (1998).

Regularization terms are almost everywhere differentiable functions $\mathcal{T}\colon \Theta \to \mathbb{R}$ depending on the parameters of the neural network architecture being trained. They are minimized next to the empirical risk $\widehat{\mathcal{R}}_{\mathcal{D}_{\text{train}}}(\theta)$, i.e., making the training algorithm minimize the quantity

$$\widehat{\mathcal{R}}_{\mathcal{D}_{\text{train}}}(\theta) + \beta\,\mathcal{T}(\theta)$$

with respect to $\theta \in \Theta$, where $\beta > 0$ is called *regularization weight* and controls the influence of the regularization term over the empirical risk.

4.2 Pruning

Modern neural networks typically consist of highly complex architectures with a large number of parameters. Working with such models usually requires large amounts of computational resources that may not be easily available or sustainable in the long term. One of the approaches to solve this problem is *neural network pruning*. Pruning involves systematically removing parameters from an existing network so that the resulting network after this process conserves as much performance as possible, but with a drastically reduced number of parameters (Blalock et al. 2020).

4.3 Adversarial, Out-of-Distribution, and Shifted Examples

Adversarial, out-of-distribution, and shifted examples are specific values in the input space of a neural network that have special properties with respect to the underlying data distribution or the way the neural network processes them. Given a machine learning model, *adversarial examples* are inputs that differ slightly from correctly classified inputs but are misclassified by the model. Differences between adversarial examples and properly classified regular examples are often imperceptible to human senses (Goodfellow et al. 2015). On the other hand, *out-of-distribution examples* are valid input values for a machine learning model that have not been sampled from the real data distribution. Similarly, *shifted examples* are examples that deviate from the real data distribution but come from samples from it, resembling the nature of adversarial examples (Lacombe et al. 2021).

Detecting input types is an important problem in deep learning. On one hand, it allows further study of the robustness and performance of a given neural network. On the other hand, it allows to detect attacks to machine learning models, both in training and in inference.

4.4 Detection of Trojaned Networks

In a Trojan attack on a machine learning model (Zheng et al. 2021), attackers create *trojaned* examples, that are normal training examples to which some extra information, using specific patterns, is added. This extra information is used by attackers to generate specific outputs in the inputs containing these specific patterns, allowing attackers to bypass the regular outputs of the model during inference. To be able to generate these *unwanted* outputs, attackers inject trojaned examples into the original training data set without the knowledge of the legitimate owner of the model before the model is trained with the objective that the network recognizes the pattern and acts according to the attackers' intention.

Detecting neural networks trained with trojaned examples is vital to avoid security breaches in machine learning, especially in critical-context applications such as medical systems or security systems, among others.

4.5 Model Selection

Over the last years, there has been an explosion on the quantity and quality of models developed by deep learning practitioners. Most of these models have a large number of parameters and are hard and expensive to train. However, once trained successfully, they generalize well to many situations. For this reason, a reasonable practice in nowadays applications is, instead of training new models from scratch, taking pre-trained models and adapting them to the particular problem the user is trying to solve.

However, selecting the pre-trained model that best fits the user's requirements from a *bank* of models is not a trivial task. For starters, each bank of models has its own models, trained in different datasets with different algorithms and different hardware. Small variations of the parameters in the same architecture can induce important differences during inference, so selecting only by the architecture is not usually the best option. Additionally, due to the growing number of options available, testing all of them may be an unfeasible process for practical applications. For this reason, a good model selection algorithm that takes into account many properties of the models is fundamental.

4.6 Prediction of Accuracy

In classification tasks, a commonly used loss function is given by

$$\mathcal{L}(\mathcal{N}, x, y) = \mathbb{1}_{\{(x,y)\,:\,x \neq y\}} (\mathcal{N}(x), y) . \tag{4.1}$$

The real risk $\mathcal{R}(\mathcal{N})$ for this loss function is the expectation that the neural network $\mathcal{N}$ misclassifies x. Correspondingly, the value $1 - \mathcal{R}(\mathcal{N})$ is the expectation that the neural network correctly classifies an input x. This value is called *accuracy*:

$$\mathrm{Acc}(\mathcal{N}) = 1 - \mathcal{R}(\mathcal{N}).$$

The higher the accuracy, the better the model according to the risk of $\mathcal{L}$. Thus, one way to study the performance of neural networks in classification tasks is to study their accuracy values.

The problem with $\mathrm{Acc}(\mathcal{N})$ is that we cannot compute it, so we approximate it by the *empirical accuracy* in a sample different from the training dataset $\mathcal{D}_{\mathrm{train}}$, called *test dataset* and denoted by $\mathcal{D}_{\mathrm{test}}$. The empirical accuracy of $\mathcal{N}$ in a sample $\mathcal{D}$ is

$$\widehat{\mathrm{Acc}}_{\mathcal{D}}(\mathcal{N}) = \frac{1}{|\mathcal{D}|} \sum_{(x,y) \in \mathcal{D}} \mathbb{1}_{\{(x,y)\,:\,x = y\}} (\mathcal{N}(x), y) . \tag{4.2}$$

Empirical accuracy on a test dataset $\mathcal{D}_{\mathrm{test}}$ is one of the most used metrics to assess the generalization and performance of a trained neural network. However, in certain situations, particularly in data-scarce scenarios, it is beneficial to have performance measures that do not rely on an additional dataset. Thus, developing measures computed from properties of neural networks that can be used to *predict* the test accuracy serves a dual purpose: it provides a performance measure independent of an extra dataset and captures the causes making the model (in)effective.

Another similar measure that provides information about the performance of a network is the (empirical) generalization gap. The *empirical generalization gap* is the difference between the accuracy in a test dataset and the accuracy in the training dataset, i.e.,

$$\widehat{\mathrm{Gap}}(\mathcal{N}) = \widehat{\mathrm{Acc}}_{\mathcal{D}_{\mathrm{test}}}(\mathcal{N}) - \widehat{\mathrm{Acc}}_{\mathcal{D}_{\mathrm{train}}}(\mathcal{N}).$$

Arguably, predicting the generalization gap is at least as useful as predicting the test accuracy of a network, as it can be used to compute the test accuracy only using the training dataset. An analysis of methods that try to predict this last measure can be found in Jiang et al. (2021). Also, a large-scale study of generalization bounds and measures of generalization in deep learning can be found in Jiang et al. (2020).

4.7 Quality Assessment of Generative Models

Neural network quality measurements in classification tasks are more straightforward to define than in generative tasks. In general, there are many interesting factors that can be evaluated to compare the quality of two generative neural networks, and thus deciding which are the best ones is a hard challenge. For example, the *Fréchet inception distance* (FID) (Heusel et al. 2017), one of the most popular quality metrics for generative models, quantifies the similarity between the original data distribution $\mathbb{P}_X$ and the distribution induced by the generator. Specifically, FID computes the Fréchet distance between two multivariate Gaussians fitted to the activations of a pre-trained neural network, typically Inception v3 (Szegedy et al. 2016), when processing real and generated data samples.

References

Davis Blalock, Jose J. Gonzalez Ortiz, Jonathan Frankle, and John Guttag. What is the state of neural network pruning? In I. Dhillon, D. Papailiopoulos, and V. Sze, editors, *Proceedings of the 3rd MLSys Conference, Austin, TX, USA*, volume 2 of *Proceedings of Machine Learning and Systems*, pages 129–146, 2020. URL https://proceedings.mlsys.org/paper_files/paper/2020/file/6c44dc73014d66ba49b28d483a8f8b0d-Paper.pdf.

Ian J. Goodfellow, Jonathon Shlens, and Christian Szegedy. Explaining and harnessing adversarial examples, 2015. URL https://arxiv.org/abs/1412.6572.

Martin Heusel, Hubert Ramsauer, Thomas Unterthiner, Bernhard Nessler, and Sepp Hochreiter. Gans trained by a two time-scale update rule converge to a local nash equilibrium. In I. Guyon, U. Von Luxburg, S. Bengio, H. Wallach, R. Fergus, S. Vishwanathan, and R. Garnett, editors, *Advances in Neural Information Processing Systems*, volume 30. Curran Associates, Inc., 2017. URL https://proceedings.neurips.cc/paper_files/paper/2017/file/8a1d694707eb0fefe65871369074926d-Paper.pdf.

Yiding Jiang, Behnam Neyshabur, Hossein Mobahi, Dilip Krishnan, and Samy Bengio. Fantastic generalization measures and where to find them. In *International Conference on Learning Representations*, 2020. URL https://openreview.net/forum?id=SJgIPJBFvH.

Yiding Jiang, Parth Natekar, Manik Sharma, Sumukh K. Aithal, Dhruva Kashyap, Natarajan Subramanyam, Carlos Lassance, Daniel M. Roy, Gintare Karolina Dziugaite, Suriya Gunasekar, Isabelle Guyon, Pierre Foret, Scott Yak, Hossein Mobahi, Behnam Neyshabur, and Samy Bengio. Methods and analysis of the first competition in predicting generalization of deep learning. In Hugo Jair Escalante and Katja Hofmann, editors, *Proceedings of the NeurIPS 2020 Competition and Demonstration Track*, volume 133 of *Proceedings of Machine Learning Research*, pages 170–190. PMLR, 06–12 Dec 2021. URL https://proceedings.mlr.press/v133/jiang21a.html.

Théo Lacombe, Yuichi Ike, Mathieu Carrière, Frédéric Chazal, Marc Glisse, and Yuhei Umeda. Topological uncertainty: Monitoring trained neural networks through persistence of activation graphs. In Zhi-Hua Zhou, editor, *Proceedings of the Thirtieth International Joint Conference on Artificial Intelligence, IJCAI-21*, pages 2666–2672. International Joint Conferences on Artificial Intelligence Organization, 8 2021. URL https://doi.org/10.24963/ijcai.2021/367. Main Track.

Lutz Prechelt. *Early Stopping – But When?*, pages 55–69. Springer Berlin Heidelberg, Berlin, Heidelberg, 1998. ISBN 978-3-540-49430-0. URL https://doi.org/10.1007/3-540-49430-8_3.

Christian Szegedy, Vincent Vanhoucke, Sergey Ioffe, Jon Shlens, and Zbigniew Wojna. Rethinking the inception architecture for computer vision. In *2016 IEEE Conference on Computer Vision and Pattern Recognition (CVPR)*, pages 2818–2826, 2016. https://doi.org/10.1109/CVPR.2016.308.

Chiyuan Zhang, Samy Bengio, Moritz Hardt, Benjamin Recht, and Oriol Vinyals. Understanding deep learning (still) requires rethinking generalization. *Commun. ACM*, 64(3):107–115, feb 2021. ISSN 0001-0782. URL https://doi.org/10.1145/3446776.

Songzhu Zheng, Yikai Zhang, Hubert Wagner, Mayank Goswami, and Chao Chen. Topological detection of trojaned neural networks. In A. Beygelzimer, Y. Dauphin, P. Liang, and J. Wortman Vaughan, editors, *Advances in Neural Information Processing Systems*, 2021. URL https://openreview.net/forum?id=1r2EannVuIA.

Chapter 5
Input and Output Spaces

Abstract In this chapter, we discuss how topological data analysis has been used to analyze the structure of neural networks, together with decision regions, boundaries, decompositions of input spaces, and input and output spaces of generative and non-generative models.

5.1 Structure of Neural Networks

Recall that a fully connected feedforward neural network $\mathcal{N}$ is specified by an architecture $a = (N, \varphi)$ and a set of parameters Θ depending on the architecture a. At the same time, recall that the architecture a defines a directed graph $G(\mathcal{N})$ that characterizes the computations of the neural network function, where each vertex is associated with some value during the evaluation of $\phi_{\mathcal{N}}$.

Thus, the structure of this graph is relevant for studying the generalization capacity and expressiveness of neural networks. Given the directed nature of the graph, it is preferable to avoid analyzing $G(\mathcal{N})$ while disregarding its directions, as is often the case with the general topological tools discussed in Chap. 3.

Chowdhury et al. (2019) circumvent the directionality problem by studying and comparing the ranks of two special homologies designed for directed graphs in the context of FCFNNs. These are *path homology* (Grigor'yan et al. 2014) and the homology of *directed flag complexes* (DFC) (Reimann et al. 2017), denoted by $\mathrm{PathHom}_k(G(\mathcal{N}))$ and $\mathrm{DFCHom}_k(G(\mathcal{N}))$, respectively. Here k denotes the homological dimension, as in the simplicial case.

In particular, Chowdhury et al. (2019) showed that, given a neural network $\mathcal{N}$ with architecture $a = (N, \varphi)$, where $N = (N_0, \ldots, N_L)$, and given a dimension $k \in \mathbb{N}$, the following equalities hold:

$$\mathrm{rank}(\mathrm{PathHom}_k(G(\mathcal{N}))) = \delta_{k,L-1} \prod_{i=0}^{L}(N_i - 1),$$

$$\mathrm{rank}(\mathrm{DFCHom}_k(G(\mathcal{N}))) = \mathrm{rank}(H_k(G(\mathcal{N}))),$$

© The Author(s), under exclusive license to Springer Nature Switzerland AG 2026

R. Ballester et al., *Topological Data Analysis for Neural Networks*, SpringerBriefs in Computer Science, https://doi.org/10.1007/978-3-032-08283-1_5

where $\delta_{k,\ell} = 1$ if $k = \ell$ and $\delta_{k,\ell} = 0$ otherwise, and $H_k(G(\mathcal{N}))$ is the k-dimensional simplicial homology group of $G(\mathcal{N})$ seen as an undirected graph. It can be further shown that

$$\mathrm{rank}(H_k(G(\mathcal{N}))) = \begin{cases} 1 & \text{if } k = 0, \\ 1 - |V(G(\mathcal{N}))| + |E(G(\mathcal{N}))| & \text{if } k = 1, \\ 0 & \text{if } k \geq 2. \end{cases}$$

The equality between the DFC-homology ranks and the standard simplicial homology ranks for neural network graphs implies that DFCHom forgets the directions of neural network graphs. Note also that the ranks for both families of homology groups can be directly computed from the number of neurons and edges contained by the network, and do not reflect most of the structural properties of the neural network graph. For example, path homologies are invariant to layer permutations, although the order of the layers is a key factor in achieving good neural network performance. Furthermore, DFC-homology forgets all the structural information except for the number of vertices and edges, which many network configurations with different performances share.

Both path and DFC-homology for neural networks have an important drawback, namely that they forget the weights of neural networks, which are crucial in their performance. For DFC-homology, this can be overcome by using usual persistent homology on the undirected neural network graphs, as homology without weights coincide. For path homology, this can be remedied by studying a *persistent* version of it, called *persistent path homology* (Chowdhury and Mémoli 2018).

To the best of our knowledge, persistent path homology has not yet been applied to the analysis of neural networks. This presents a promising direction for future research, particularly in light of the recent development of a more efficient algorithm for computing path homology (Zhu and Chi 2024).

5.2   General Output Space and Decision Regions

Decision regions and boundaries are crucial for understanding the predictions of a neural network in a classification problem, as they delimit the regions of the input space $\mathcal{X}$ associated to each label and the regions where the neural network is *uncertain* about the label. Decision regions and boundaries reflect important properties of the neural network, such as its generalization capacity (Guan and Loew 2020) or its robustness against adversarial perturbations (Moosavi-Dezfooli et al. 2019).

One of the first articles in this survey that studied decision regions using topological data analysis was published by Bianchini and Scarselli (2014). In it, the sum of Betti numbers of the decision regions of one of the labels of a binary classification problem was analyzed and theoretically bounded for neural networks with a special focus on neural networks implementing Pfaffian activations (Khovanskiĭ 1991). A difficulty in the previous paper is that bounds of Betti numbers are not tight and cannot be used easily in practical scenarios.

Building upon this foundation, Guss and Salakhutdinov (2018) took an experimental approach to further explore the Betti numbers of decision regions. They empirically studied such Betti numbers for several trained neural networks and compared them with the Betti numbers of the real region induced by the selected label.

On a more general setting, Liu et al. (2023a) proposed the GTDA algorithm of Sect. 3.2 as an improved version of Mapper to study the input space based on the output values of general neural network functions, with the aim of enhancing the interpretability of the decisions of neural networks. They studied Reeb networks built using GTDA with input graphs induced from a dataset $\mathcal{D}$ containing one node per sample x in $\mathcal{D}$. The filter functions used in their work were given by

$$\left(\phi_N^{(L)}(x)_1, \ldots, \phi_N^{(L)}(x)_{N_L}\right),$$

for each $x \in \mathcal{D}$ plus, possibly, some extra values depending on the dataset or the outputs. To induce the edges from $\mathcal{D}$, they either used previously-known binary relationships between the data or discovered binary relationships given by algorithms such as the nearest neighbors algorithm of the raw or transformed data.

Liu et al. (2023a) successfully extracted valuable insights from GTDA graphs about the Enformer model predictions (Avsec et al. 2021) on a DNA dataset that Mapper and other alternative methods were not able to capture. They also employed GTDA graphs for automatic error estimation in machine learning prediction tasks. By comparing predicted labels with ground truth labels of nearby samples according to a proximity notion based on distances in the GTDA graphs, they calculated numerical error estimations that successfully corrected erroneous predictions in binary classification tasks and identified regions with high misclassification rates.

Further experiments extended this approach to other network architectures and datasets, including ResNet-50 and Imagenette (Howard 2021). These experiments yielded positive results, validating GTDA as a powerful tool for deep learning analysis and interpretability.

5.3   Decision Boundaries of a Single Network

Decision boundaries, defined in (2.4), are relevant in classification problems, as they determine the decision regions of the neural network and the spaces in which neural network decisions have more than one valid output. Unless otherwise stated, we use classification neural networks with $N_L = k$ and $\pi(x_1, \ldots, x_k) = \mathrm{argmax}_{i \in [k]} x_i$, as explained in Sect. 2.4.

Ramamurthy et al. (2019) proposed a modified version of Čech simplicial complexes, which, given a large enough sample of points near the decision region of a binary classification neural network, can reconstruct a space that is homotopy equivalent to the real decision boundary induced by the neural network under some mild assumptions about the decision boundary.

However, Čech complexes are too costly to compute in practical scenarios. For this reason, they propose two computationally feasible variants of the Vietoris–Rips filtrations and complexes that aim to recover the Betti numbers of the decision boundaries of neural networks using persistence diagrams induced by the new filtrations.

These new filtrations, denoted by $(\text{P-LVR}_t(\mathcal{D}))_{t\in\mathbb{R}}$ and $(\text{LS-LVR}_t(\mathcal{D}))_{t\in\mathbb{R}}$, and called *plain LVR* and *locally scaled LVR*, respectively, are constructed from a sample $\mathcal{D} = \{(x_1, y_1), \ldots, (x_m, y_m)\}$ containing elements from both classes. Specifically, given $\bullet \in \{\text{P-LVR}, \text{LS-LVR}\}$, the simplicial complexes of these filtrations are

$$\bullet_t (\mathcal{D}) = \{\sigma \subseteq \{x_i\}_{i=1}^m : \{\sigma_i, \sigma_j\} \in G_t^\bullet(\mathcal{D}) \text{ for all } \{\sigma_i, \sigma_j\} \subseteq \sigma\},$$

where $G_t^\bullet(\mathcal{D})$ is the 1-dimensional simplicial complex consisting of the vertices and edges of a bipartite graph $G_t^\bullet = (V_1, V_2, E_t)$ with $V_j = \{x_i : y_i = j\}$ for $j \in \{1, 2\}$ satisfying $\{v_1, v_2\} \in E_t$ if and only if $d_\bullet(v_1, v_2) \leq t$, plus a set of new edges joining the vertices connected by paths of length two in $G_t^\bullet$. The dissimilarities $d_\bullet$ are given by

$$d_{\text{P-LVR}}(v_1, v_2) = \|v_1 - v_2\|_2, \qquad d_{\text{LS-LVR}}(v_1, v_2) = \frac{\|v_1 - v_2\|_2}{\sqrt{\rho_1 \rho_2}},$$

where $\rho_i, i \in \{1, 2\}$ is the local scale of v_i, that is, the radius of the smallest sphere centered at v_i that encloses at least k points from the opposite class, where k is a fixed parameter.

A problem with the previous constructions is that they may require very large samples to robustly recover the topology of the decision boundaries. Obtaining such samples can sometimes be infeasible, especially when the cost of obtaining the value of $y \in \mathcal{Y}$ given $x \in \mathcal{X}$ is high. For example, in a medical binary classification problem, given a radiography as an image $x \in \mathcal{X}$, diagnosing whether there is a problem with the patient ($2 \in \mathcal{Y}$) or not ($1 \in \mathcal{Y}$) requires a specialist investing time on analyzing the image, which can be expensive in time and resources.

Li et al. (2020) proposed a method based on active learning (Settles 2009) to sample a small but meaningful set to recover the homology of the decision boundaries using the labeled Čech and Vietoris–Rips complexes introduced in the previous article. Persistence diagrams were used in both articles to perform model selection for basic datasets including Banknote (Lohweg 2013), CIFAR-10 (Krizhevsky 2009), MNIST (Deng 2012), and Fashion-MNIST (Xiao et al. 2017), and basic machine and deep learning models.

Decision boundaries are directly related to the performance of neural networks. Given a binary classification task and a neural network $\mathcal{N}$ with $N_L = 1$ and projection function $\pi(x) = 2$ if $x > 0$ and $x \le 0$, recall that the decision boundary of $\mathcal{N}$ is the set $\phi_{\mathcal{N}}^{-1}(0)$, as in Eq. (2.5) for $b = 0$. Let $x \in \mathrm{Dom}(\phi_{\mathcal{N}})$. For this kind of neural network, the higher the absolute value of $\phi_{\mathcal{N}}(x)$, the more likely it is that the same input, slightly perturbed, is classified with the same label as x, that is, $\pi(\phi_{\mathcal{N}}(x)) = \pi(\phi_{\mathcal{N}}(x + \varepsilon))$, for a small ε. In other words, the higher the absolute value of $\phi_{\mathcal{N}}$ for a given example x, the more robust the prediction of $\mathcal{N}$ for x.

Chen et al. (2019) propose a regularization term to remove *weak* connected components from the decision boundary, that is, connected components enclosing decision regions such that the predictions for their points are not robust according to the high absolute value criterion. This regularization term for binary classification tasks can be computed from zero-dimensional zigzag persistence diagrams induced by $\phi_{\mathcal{N}}$ whenever the neural network function $\phi_{\mathcal{N}}$ is Morse, which occurs in several neural network architectures (Kurochkin 2021), and $\mathcal{X}$ is a hypercube. This regularization term was the first to use the differentiability properties of persistent homology, studied and introduced in parallel by Leygonie et al. (2022) and by Carrière et al. (2021). The effectivity of the term was studied for a simple kernel logistic regression, that was compared to other non-deep learning models such as KNN, logistic regression, or SVM, among others, trained with L^1 and L^2 regularization terms. The experiments were performed in synthetic and real datasets, where the real datasets came from the UCI dataset bank (Kelly et al. 1987) and from two biomedical datasets (Yuan et al. 2014; Ni et al. 2018). In most cases, the model trained with the topological regularizer outperformed the rest of the models. However, no deep learning models were involved, and further experimentation is needed to see the efficacy of this approach for neural networks.

5.4 Decision Boundaries of a Hypothesis Set

Earlier articles studied decision regions and boundaries for specific instances of neural networks, that is, pairs consisting of an architecture and a particular set of parameters. Petri and Leitão (2020) analyzed the topology of all possible decision boundaries given by a hypothesis set $\mathcal{F}_{a,\Theta}$ for a given fixed architecture a and a given fixed set of possible parameters Θ. The objective of this study was to associate a *topological diversity measure* to a particular hypothesis set $\mathcal{F}_{a,\Theta}$ that characterizes the diversity in the topology of the different decision boundaries induced by the hypothesis set.

To do this, a diversity measure is computed on the metric space of persistence diagrams derived from approximations of the decision boundaries of neural networks in the hypothesis set. The procedure is as follows. First, a sample is taken with a large number of possible parameters $\{\theta_1, \ldots, \theta_m\} \subseteq \Theta$. Then, for each parameter θ_i, $i \in \{1, \ldots, m\}$, zero- and one-dimensional Vietoris–Rips persistence diagrams $D(\mathbb{V}_k(\mathrm{VR}(P_i, \|\cdot\|_2)))$ of a sample of points P_i from an approximation of the decision boundary of the neural network with architecture a and parameters θ_i are computed. Finally, topological diversity measures of the hypothesis set are given by the spread of the metric spaces (Willerton 2015) whose points are the computed persistence diagrams for a fixed dimension and whose distance is a q-Wasserstein distance for a fixed $q \in \mathbb{N}$.

To approximate spreads, which are computationally expensive to compute, Petri and Leitão used the averages of the quantities $C(D) = W_1^1(D, \Delta)$, where Δ is the diagonal in a persistence diagram, calculated from the previously computed persistence diagrams because they saw that the quantities $C(D)$ correlate with their spreads and thus also characterize the topological diversity of the hypothesis set, while being easier to compute.

5.5 Input Space Decomposition into Convex Polyhedra

Neural networks with a ReLU activation function $\varphi(x) = \max(0, x)$ determine a decomposition of the input space into convex polyhedra that assigns to each polyhedron in the decomposition a unique binary vector in such a way that two polyhedra share a facet if and only if their associated binary vectors differ exactly in one component. Liu et al. (2023b) used this polyhedral decomposition of the input space to infer the topology of manifolds embedded in the domain of a given ReLU neural network from a sample of their points using persistence diagrams.

To compute persistence diagrams given a sample $S = \{x_i\}_{i=1}^m$ of a manifold $\mathbb{M}$, the given procedure computes the set of unique binary vectors $\mathfrak{B}(S) = \{b(x_i)\}_{i=1}^m$ associated with the points of the sample, where $b(x_i)$ is the binary vector associated with the convex polyhedra in which x_i lies, and then computes the Vietoris–Rips persistence diagrams $D(\mathbb{V}_k(\mathrm{VR}(\mathfrak{B}(S), h)))$ of $\mathfrak{B}(S)$ using the Hamming distance h, which counts the number of different bits in the two binary strings to compare. Although this is not directly related to neural network analysis, we firmly believe that this opens a way to study the structure and topology of the unknown data distribution for a specific learning problem by applying this method to large enough data samples.

Polyhedral decompositions of neural networks can be seen as polyhedral complexes, a generalization of simplicial complexes by means of polyhedra. In more detail, polyhedral complexes are sets of polyhedra such that every face of a polyhedron of the complex is also in the complex, and the intersection of any two polyhedra is either empty or a face of both. For neural networks with ReLU activations, decision boundaries can be seen as subcomplexes of the polyhedral decompositions of neural networks. Masden (2022) used this fact to study the topology of one-point compactifications of decision boundaries of small, non-trained, ReLU neural networks.

Masden (2022) discovered that, under certain technical conditions, polyhedral decompositions possess dual structures known as *signed sequence cubical complexes* (SSCCs), which are specific types of cubical complexes. Like polyhedral complexes, cubical complexes offer an alternative to simplicial complexes. The key distinction is that cubical complexes use cubes instead of simplices and are generally more manageable than polyhedral complexes. SSCCs allow us to define mod-2 coboundary maps on the one-point compactification of polyhedral decompositions. These maps are dual to the mod-2 boundary maps on cubical complexes, and can also be used by restriction on the one-point compactification of the decision boundary from which Masden (2022) obtained (co)homology groups.

Differences were found in the topology of decision boundaries in terms of their neural network layer structures. In networks with a single hidden layer, the Betti numbers of decision boundaries exhibited remarkable consistency across various parameter ranges, including modifications to input data dimension and the number of hidden neurons. In contrast, networks with two hidden layers displayed more variability in the distribution of the Betti numbers of their decision boundaries.

5.6 Generative Neural Networks

TDA has also been used in generative deep learning. One of the most well-known class of models for generative deep learning are generative adversarial networks (GAN) (Goodfellow et al. 2014). Generative adversarial networks try to generate synthetic samples from a *real* data distribution living in an ambient space X distributed by an unknown function $\mathbb{P}_X$.

To do this, generative adversarial models are composed of two neural networks: the *generator* G and the *discriminator* S, equipped with functions $\phi_G \colon Z \to X$ and $\phi_S \colon X \to \{1, 2\}$, respectively, where Z is a space distributed by a known distribution $\mathbb{P}_Z$. We refer to Z as the *noise space* of the generative adversarial network. The generator is a neural network whose objective is to generate synthetic samples from samples from the noise space. The discriminator is a network that, given a sample —generated by G or sampled from the real data distribution—, tries to distinguish between the two.

The training of a generative adversarial network is performed by training the generator and the discriminator in an adversarial way, that is, the generator tries to fool the discriminator and the discriminator tries to distinguish between the synthetic and the real samples.

A significant challenge in the field of generative models lies in developing robust evaluation metrics. Quantifying the quality of these models remains a complex task. To this end, Khrulkov and Oseledets (2018) proposed the *geometry score*, a persistent homology-based quality metric for generative adversarial networks measuring the difference between the topology of real and synthetic samples. This metric is based on the following three assumptions:

1. The higher the quality of the generator, the more similar are the spaces $\mathrm{supp}(\mathbb{P}_X)$ and $\phi_G(\mathrm{supp}(\mathbb{P}_Z))$;
2. $\mathrm{supp}(\mathbb{P}_X)$ is concentrated in a manifold $\mathcal{M}_X$ and $\phi_G(\mathrm{supp}(\mathbb{P}_Z))$ is a manifold;
3. The higher the similarity between the spaces $\mathcal{M}_X$ and $\phi_G(\mathrm{supp}(\mathbb{P}_Z))$, the higher the similarity between their topologies.

Note that the first part of the second assumption is the manifold hypothesis, and it may not hold in general scenarios (Von Rohrscheidt and Rieck 2023). However, this limitation does not affect the computation of persistent homology of point clouds, which is the basis for the metric. Consequently, the metric should be used with care because, although its value can always be computed, it may be meaningless if assumption 2 is not satisfied.

To compare both spaces $\mathcal{M}_X$ and $\phi_{\mathcal{G}}(\mathrm{supp}(\mathbb{P}_Z))$, Khrulkov and Oseledets (2018) computed persistence modules of witness filtrations (de Silva and Carlsson 2004) for samples from X and $\phi_{\mathcal{G}}(\mathcal{Z})$, asuming that the samples taken from X and from $\phi_{\mathcal{G}}(\mathcal{Z})$ are also contained in $\mathcal{M}_X$ and $\phi_{\mathcal{G}}(\mathrm{supp}(\mathbb{P}_Z))$, respectively. Witness filtrations have the advantage that they are usually efficient even for big point clouds, although they may be more challenging to use effectively. Given a point cloud (P, d), a subset of *landmarks* $L \subseteq P$, and $\alpha \geq 0$, the α-*witness complex* of (P, d) with landmarks L is the simplicial complex

$$W(P, L, \alpha, d) = \left\{ \sigma \subseteq L : \exists w \in X \; \forall l \in \sigma \; \forall l' \in L \smallsetminus \sigma, \right.$$

$$\left. d(w, l)^2 \leq d(w, l')^2 + \alpha \right\}. \qquad (5.1)$$

Choosing $\alpha_{\max} > 0$, we have that $W_{\alpha_{\max}} = (W(P, L, \alpha, d))_{\alpha \in [0, \alpha_{\max}]}$ is a filtration of simplicial complexes that induces a persistence module $\mathbb{V}_k(W_{\alpha_{\max}})$ where, for $t > \alpha_{\max}$, all the vector spaces forming the persistence module are equal to the one corresponding to the value $t = \alpha_{\max}$.

To measure the difference between two persistence modules induced by the spaces $\mathcal{M}_X$ and $\phi_{\mathcal{G}}(\mathrm{supp}(\mathbb{P}_Z))$ using witness filtrations, Khrulkov and Oseledets (2018) introduced the *relative living time* RLT of i-features for k-dimensional homology, given by

$$\mathrm{RLT}(i, k, P, L, d) = \frac{\mu\big(\{\alpha \in [0, \alpha_{\max}] : \beta_k(W(P, L, \alpha, d)) = i\} \big)}{\alpha_{\max}}. \qquad (5.2)$$

However, such persistence modules, and thus the relative living times, are inherently dependent on a choice of landmarks, making RLTs unreliable to be the basis of the geometry score. Instead, to mitigate this dependency, Khrulkov and Oseledets (2018) proposed to compare the topology of $\mathcal{M}_X$ and $\phi_{\mathcal{G}}(\mathrm{supp}(\mathbb{P}_Z))$ using *mean relative living times* (MLRTs), defined as expectations of relative living times over all possible choices of landmarks L of a fixed size, i.e.,

$$\mathrm{MRLT}(i, k, P, d) = \mathbb{E}_L[\mathrm{RLT}(i, k, P, L, d)].$$

Building upon MRLTs, the geometry score of a generative adversarial network $\mathcal{G}$ was defined by means of the squared differences of all the MRLTs, for all i, for the 1-dimensional persistence modules generated by samples P and P' equipped with the same dissimilarity function d from $\mathcal{X}$ and $\phi_{\mathcal{G}}(\text{supp}(\mathbb{P}_Z))$, respectively, that is,

$$\sum_{i=0}^{\infty} \big(\text{MLRT}(i, 1, P, d) - \text{MLRT}(i, 1, P', d)\big)^2.$$

The validity of the geometry score as a quality metric was proved in several experiments. For example, it was ratified that the WGAN-GP (Gulrajani et al. 2017) is better than the WGAN model (Arjovsky et al. 2017), a well-known result, for the MNIST dataset. Additionally, the geometry score was able to distinguish between good and bad generative models derived from the DCGAN architecture (Radford et al. 2016) trained on the CelebA dataset (Liu et al. 2015).

In a very similar fashion, but with a simpler approach, Charlier et al. (2019) proposed to compare generative models, this time not necessarily restricted to GAN models, by directly comparing the zero and one-dimensional persistence diagrams induced by Vietoris–Rips filtrations of samples taken from the real data distribution and from the generative model using the bottleneck distance. In this case, the experiments compared four different generative models in the credit card fraud detection dataset (Credit Dataset). These models were WGAN and WGAN-GP together with WAE (Tolstikhin et al. 2018) and VAE (Kingma and Welling 2014). In this case too, WGAN-GP was shown to produce better results than the other models according to the persistent homology results.

The two aforementioned approaches evaluating the quality of generative models were based on comparing the topology of real and synthetic data by means of persistent homology. However, they are not the usual choices in the generative deep learning community, where scores such as the Fréchet inception distance (Heusel et al. 2017) or a numerical approximation of precision and recall (Sajjadi et al. 2018) are currently preferred to evaluate generative models.

Kim et al. (2023) proposed an alternative way to approximate precision and recall scores using an approximation of the support of the data distributions based on preimages of kernel density estimators (KDEs). Their approximation was built upon the results of Fasy et al. (2014, Method IV), where it is argued that persistence diagrams from superlevel set persistence modules may carry topological information from the support of the data distribution. Specifically, the supports of the real and synthetic data were approximated by the superlevel sets $(\widehat{p}_{h_r})^{-1}[c_r, \infty)$ and $(\widehat{p}_{h_s})^{-1}[c_s, \infty)$, respectively, where $\widehat{p}_{h_r}$ and $\widehat{p}_{h_s}$ are KDEs for real and synthetic datasets $\mathcal{D}_r$ and $\mathcal{D}_s$ given by

$$\widehat{p}_{h_\bullet}(x) = \frac{1}{|\mathcal{D}_\bullet|} \sum_{p \in \mathcal{D}_\bullet} \frac{1}{h^d} K\left(\frac{x - p}{h}\right),$$

where $h > 0$ and K are the bandwidth and kernel of the KDEs, selected beforehand, and c_r and c_s are confidence bands for a given significance value α, obtained using bootstrap over the datasets $\mathcal{D}_r$ and $\mathcal{D}_s$ with the condition that

$$\liminf_{|\mathcal{D}_\bullet| \to \infty} \ \mathbb{P}\big(\|\widehat{p}_{h_\bullet} - p_{h_\bullet}\|_\infty < c_\bullet\big) \geq 1 - \alpha,$$

where p_{h_r} and p_{h_s} are smoothed versions of the real and synthetic data distributions, respectively (Fasy et al. 2014, Equation (26)). By the stability theorems (Cohen-Steiner et al. 2007), the confidence bands c_r and c_s, also bound the distances between the persistence diagrams coming from the functions $\widehat{p}_{h_\bullet}$ and $p_{h_\bullet}$ with probability $1 - \alpha$ and equip the persistence diagrams induced by $\widehat{p}_{h_\bullet}$ with a notion of *noisy points* (Fasy et al. 2014, Section 4).

The superlevel sets $(\widehat{p}_{h_\bullet})^{-1}[c_\bullet, \infty)$ are thus regions which intend to induce persistence diagrams without noisy points according to the previous confidence bands, thus recovering supports with similar topology to the topology of the support of the smoothed data distributions measured by the distances between their respective persistence diagrams.

Under some technical assumptions, the precision and recall scores approximated using the superlevel sets proposed by Kim et al. (2023) become close to the real precision and recall from the distribution as more examples are added to the datasets $\mathcal{D}_r$ and $\mathcal{D}_s$. The approximation is done with robustness, meaning that it holds even with data possibly corrupted by noise.

The suitability of the new metrics was tested on several synthetic and real scenarios, where the metric accurately described the differences between real and synthetic data distributions in several scenarios where other metrics struggle to evaluate differences. Also, the F1-score computed from the precision and recall approximations ranked different generative models such as StyleGAN2 (Karras et al. 2020), ReACGAN (Kang et al. 2021), among others, as the FID metric, making the F1-score consistent with the state-of-the-art knowledge about the quality of generative models.

GANs belong to a broader category of generative models that produce synthetic data by mapping values from a *latent space* $\mathcal{Z}$ to the ambient space $\mathcal{X}$ of real data through a function $G \colon \mathcal{Z} \to \mathcal{X}$. For such generative models, a desirable property is to enable us to control the attributes of the generated samples.

One approach to controlling attributes involves designing generative models that map from the latent space to the data space in such a way that the latent space $\mathcal{Z}$ can be factored into subspaces corresponding to *factors of variation* in the generated data. In other words, when generating data from samples $z_1, \ldots, z_m \in \mathcal{Z}$ where all components are fixed except those corresponding to a single factor of variation, the resulting generated data $G(z_1), \ldots, G(z_m)$ vary only along that specific source of variation. For example, if we are generating rectangle shapes in $\mathbb{R}^2$ where the sides are parallel to the x, y axes from a latent space $\mathcal{Z}$, we would want to be able to factor $\mathcal{Z}$ into $\mathcal{Z}_1 \times \mathcal{Z}_2 \times \mathcal{Z}_3 \times \mathcal{Z}_4$ where the components of (z_1, z_2, z_3, z_4) control the lengths of the vertical and horizontal sides, and the vertical and horizontal distance from the origin, respectively.

A model that satisfies this property is said to be *disentangled.* For a formal discussion of disentanglement, we refer the reader to Higgins et al. (2018). Preliminary work on disentangling generative models using topological data analysis to build regularization terms can be found in Balabin et al. (2023). In their work, regularization terms minimize the topological difference between generated data from two samples of the latent space forced to belong to the same factor of variation using the representation topology divergence, a method to compare point clouds topologically which is reviewed in Sect. 6.2.

Measuring a model's entanglement is challenging, and there is no canonical way to do it. Zhou et al. (2021) proposed two disentanglement measures based on topological data analysis: one unsupervised, that can be computed without the need of a real dataset, and one supervised, that uses a real data reference. These measures demonstrated comparable performance to other existing metrics in different datasets and architectures, while requiring fewer assumptions about the dataset or model. For both metrics, the main assumptions are that $\mathrm{Im}(G)$ is a manifold and that the images of G, when restricted to a fixed value for one factor of variation, form submanifolds of $\mathrm{Im}(G)$. In the supervised case, the only additional assumption is that the data lies on a manifold. The proposed entanglement measures are based on two fundamental ideas for disentangled models:

1. Submanifolds corresponding to fixing values for different factors of variation are usually not homeomorphic;
2. For a given factor of variation, the submanifolds of $\mathrm{Im}(G)$ generated by G fixing different values of the factor are all homeomorphic and therefore share the same topological structure.

The proposed measures penalize inconsistent topologies among submanifolds derived from varying the same factor of variation while rewarding topological differences between submanifolds obtained by fixing values of different factors. However, a significant challenge arises from the fact that the association of the latent dimensions of $\mathcal{Z}$ and the factors of variation is often unknown, making it difficult to obtain the submanifolds corresponding to specific factors of variation.

Technical limitations like the previous one preclude the computation of full submanifolds for each restriction, and thus, the comparison of classical topological measures between them. To address this issue, the topological differences between submanifolds of different factors of variation are approximated by the topological differences of the submanifolds induced by the latent variables. Specifically, the submanifolds are obtained by fixing values for the latent variables instead of using factors of variation, and the topology for each submanifold is characterized by relative living times, that only need samples from the submanifolds.

Then, Wasserstein barycenters (Agueh and Carlier 2011) are calculated for each latent dimension from the relative living times, leading to a dissimilarity matrix M based on the pairwise Wasserstein distances between the Wasserstein barycenters. This matrix, processed by a (co)clustering algorithm, produces another dissimilarity matrix M' measuring the *topological similarities* of c different probable factors of variation, from which the unsupervised disentanglement measure is derived as

$$\mu = \mathrm{tr}(M') - \left(\sum_{i=1}^{c} \sum_{j=1}^{c} M'_{i,j} - \mathrm{tr}(M') \right). \tag{5.3}$$

The supervised version, assuming that the data space is already factored into subspaces corresponding to the factors of variation, differs by generating the matrix M, not necessarily squared, as a Wasserstein dissimilarity matrix between the Wasserstein barycenters of the latent variables and of the factored real data. Finally, the supervised metric is calculated as in (5.3) for the leading principal submatrix of M', generated from M as before, taking the first c rows and columns.

References

Martial Agueh and Guillaume Carlier. Barycenters in the Wasserstein space. *SIAM Journal on Mathematical Analysis*, 43(2):904–924, 2011. doi: 10.1137/100805741. URL https://doi.org/10.1137/100805741.

Martin Arjovsky, Soumith Chintala, and Léon Bottou. Wasserstein generative adversarial networks. In Doina Precup and Yee Whye Teh, editors, *Proceedings of the 34th International Conference on Machine Learning*, volume 70 of *Proceedings of Machine Learning Research*, pages 214–223. PMLR, Aug 2017. URL https://proceedings.mlr.press/v70/arjovsky17a.html.

Žiga Avsec, Vikram Agarwal, Daniel Visentin, Joseph R. Ledsam, Agnieszka Grabska-Barwinska, Kyle R. Taylor, Yannis Assael, John Jumper, Pushmeet Kohli, and David R. Kelley. Effective gene expression prediction from sequence by integrating long-range interactions. *Nature Methods*, 18(10):1196–1203, Oct 2021. ISSN 1548-7105. URL https://doi.org/10.1038/s41592-021-01252-x.

Nikita Balabin, Daria Voronkova, Ilya Trofimov, Evgeny Burnaev, and Serguei Barannikov. Disentanglement learning via topology, 2023. URL https://arxiv.org/abs/2308.12696.

Monica Bianchini and Franco Scarselli. On the complexity of neural network classifiers: A comparison between shallow and deep architectures. *IEEE Transactions on Neural Networks and Learning Systems*, 25(8):1553–1565, 2014. https://doi.org/10.1109/TNNLS.2013.2293637.

Mathieu Carrière, Frédéric Chazal, Marc Glisse, Yuichi Ike, Hariprasad Kannan, and Yuhei Umeda. Optimizing persistent homology based functions. In Marina Meila and Tong Zhang, editors, *Proceedings of the 38th International Conference on Machine Learning*, volume 139 of *Proceedings of Machine Learning Research*, pages 1294–1303. PMLR, 18–24 Jul 2021. URL https://proceedings.mlr.press/v139/carriere21a.html.

Jeremy Charlier, Radu State, et al. PHom-GeM: Persistent homology for generative models. In *The 6th Swiss Conference on Data Science (SDS), 2019 IEEE International Conference*. IEEE, 2019.

Chao Chen, Xiuyan Ni, Qinxun Bai, and Yusu Wang. A topological regularizer for classifiers via persistent homology. In Kamalika Chaudhuri and Masashi Sugiyama, editors, *Proceedings of the Twenty-Second International Conference on Artificial Intelligence and Statistics*, volume 89 of *Proceedings of Machine Learning Research*, pages 2573–2582. PMLR, 16–18 Apr 2019. URL https://proceedings.mlr.press/v89/chen19g.html.

Samir Chowdhury and Facundo Mémoli. Persistent path homology of directed networks. In *Proceedings of the Twenty-Ninth Annual ACM-SIAM Symposium on Discrete Algorithms*, SODA 2018, page 1152–1169, USA, 2018. Society for Industrial and Applied Mathematics. ISBN 9781611975031.

Samir Chowdhury, Thomas Gebhart, Steve Huntsman, and Matvey Yutin. Path homologies of deep feedforward networks. In *2019 18th IEEE International Conference on Machine Learning and Applications (ICMLA)*, pages 1077–1082, 2019. https://doi.org/10.1109/ICMLA.2019.00181.

David Cohen-Steiner, Herbert Edelsbrunner, and John Harer. Stability of persistence diagrams. *Discrete & Computational Geometry*, 37(1):103–120, Jan 2007. ISSN 1432-0444. URL https://doi.org/10.1007/s00454-006-1276-5.

Credit Dataset. Credit Card Fraud Detection: Anonymized credit card transactions labeled as fraudulent or genuine. https://www.kaggle.com/datasets/mlg-ulb/creditcardfraud/data, 2016. Accessed: 2023-12-01.

Vin de Silva and Gunnar Carlsson. Topological estimation using witness complexes. In Markus Gross, Hanspeter Pfister, Marc Alexa, and Szymon Rusinkiewicz, editors, *SPBG'04 Symposium on Point-Based Graphics 2004*. The Eurographics Association, 2004. ISBN 3-905673-09-6. https://doi.org/10.2312/SPBG/SPBG04/157-166.

Li Deng. The MNIST database of handwritten digit images for machine learning research [best of the web]. *IEEE Signal Processing Magazine*, 29(6):141–142, 2012. https://doi.org/10.1109/MSP.2012.2211477.

Brittany Terese Fasy, Fabrizio Lecci, Alessandro Rinaldo, Larry Wasserman, Sivaraman Balakrishnan, and Aarti Singh. Confidence sets for persistence diagrams. *The Annals of Statistics*, 42(6):2301 – 2339, 2014. URL https://doi.org/10.1214/14-AOS1252.

Ian Goodfellow, Jean Pouget-Abadie, Mehdi Mirza, Bing Xu, David Warde-Farley, Sherjil Ozair, Aaron Courville, and Yoshua Bengio. Generative adversarial nets. In *Advances in Neural Information Processing Systems*, pages 2672–2680, 2014. URL http://papers.nips.cc/paper/5423-generative-adversarial-nets.pdf.

Alexander Grigor'yan, Yong Lin, Yuri Muranov, and Shing-Tung Yau. Homotopy theory for digraphs, 2014. URL https://arxiv.org/abs/1407.0234.

Shuyue Guan and Murray Loew. Analysis of generalizability of deep neural networks based on the complexity of decision boundary. In *2020 19th IEEE International Conference on Machine Learning and Applications (ICMLA)*, pages 101–106, 2020. https://doi.org/10.1109/ICMLA51294.2020.00025.

Ishaan Gulrajani, Faruk Ahmed, Martin Arjovsky, Vincent Dumoulin, and Aaron Courville. Improved training of Wasserstein GANs. In *Proceedings of the 31st International Conference on Neural Information Processing Systems*, NIPS'17, page 5769–5779, Red Hook, NY, USA, 2017. Curran Associates Inc. ISBN 9781510860964.

William H. Guss and Ruslan Salakhutdinov. On characterizing the capacity of neural networks using algebraic topology, 2018. URL https://openreview.net/forum?id=H111AfbCW.

Martin Heusel, Hubert Ramsauer, Thomas Unterthiner, Bernhard Nessler, and Sepp Hochreiter. Gans trained by a two time-scale update rule converge to a local nash equilibrium. In I. Guyon, U. Von Luxburg, S. Bengio, H. Wallach, R. Fergus, S. Vishwanathan, and R. Garnett, editors, *Advances in Neural Information Processing Systems*, volume 30. Curran Associates, Inc., 2017. URL https://proceedings.neurips.cc/paper_files/paper/2017/file/8a1d694707eb0fefe65871369074926d-Paper.pdf.

Irina Higgins, David Amos, David Pfau, Sebastien Racaniere, Loic Matthey, Danilo Rezende, and Alexander Lerchner. Towards a definition of disentangled representations, 2018. URL https://arxiv.org/abs/1812.02230.

Jeremy Howard. Imagenette Dataset. https://github.com/fastai/imagenette, 2021.

Minguk Kang, Woohyeon Joseph Shim, Minsu Cho, and Jaesik Park. Rebooting ACGAN: Auxiliary classifier GANs with stable training. In A. Beygelzimer, Y. Dauphin, P. Liang, and J. Wortman Vaughan, editors, *Advances in Neural Information Processing Systems*, 2021. URL https://openreview.net/forum?id=r7UC-b67YkO.

Tero Karras, Samuli Laine, Miika Aittala, Janne Hellsten, Jaakko Lehtinen, and Timo Aila. Analyzing and improving the image quality of StyleGAN. In *2020 IEEE/CVF Conference on Computer Vision and Pattern Recognition (CVPR)*, pages 8107–8116, Los Alamitos, CA, USA, jun 2020. IEEE Computer Society. doi: 10.1109/CVPR42600.2020.00813. URL https://doi.ieeecomputersociety.org/10.1109/CVPR42600.2020.00813.

Markelle Kelly, Rachel Longjohn, and Kolby Nottingham. The UCI Machine Learning Repository. https://archive.ics.uci.edu, 1987.

Askold Khovanskiĭ. *Fewnomials*. Translations of Mathematical Monographs, 88. American Mathematical Society, Providence, 1991. ISBN 0821845470.

Valentin Khrulkov and Ivan Oseledets. Geometry score: A method for comparing generative adversarial networks. In Jennifer Dy and Andreas Krause, editors, *Proceedings of the 35th International Conference on Machine Learning*, volume 80 of *Proceedings of Machine Learning Research*, pages 2621–2629. PMLR, 10–15 Jul 2018. URL https://proceedings.mlr. press/v80/khrulkov18a.html.

Pum Jun Kim, Yoojin Jang, Jisu Kim, and Jaejun Yoo. TopP&R: Robust support estimation approach for evaluating fidelity and diversity in generative models. In *Thirty-seventh Conference on Neural Information Processing Systems*, 2023. URL https://openreview.net/ forum?id=2gUCMr6fDY.

Diederik P. Kingma and Max Welling. Auto-encoding variational Bayes. In Yoshua Bengio and Yann LeCun, editors, *2nd International Conference on Learning Representations, ICLR 2014, Banff, AB, Canada, April 14–16, 2014, Conference Track Proceedings*, 2014. URL http://arxiv. org/abs/1312.6114.

Alex Krizhevsky. Learning multiple layers of features from tiny images, 2009. URL http://www. cs.toronto.edu/~kriz/cifar.html.

Sergey V. Kurochkin. Neural network with smooth activation functions and without bottlenecks is almost surely a Morse function. *Computational Mathematics and Mathematical Physics*, 61(7): 1162–1168, Jul 2021. ISSN 1555-6662. URL https://doi.org/10.1134/S0965542521070101.

Jacob Leygonie, Steve Oudot, and Ulrike Tillmann. A framework for differential calculus on persistence barcodes. *Foundations of Computational Mathematics*, 22(4):1069–1131, Aug 2022. ISSN 1615-3383. URL https://doi.org/10.1007/s10208-021-09522-y.

Weizhi Li, Gautam Dasarathy, Karthikeyan Natesan Ramamurthy, and Visar Berisha. Finding the homology of decision boundaries with active learning. In Hugo Larochelle, Marc'Aurelio Ranzato, Raia Hadsell, Maria-Florina Balcan, and Hsuan-Tien Lin, editors, *Advances in Neural Information Processing Systems*, volume 33, pages 8355–8365. Curran Associates, Inc., 2020. URL https://proceedings.neurips.cc/paper_files/paper/2020/file/ 5f14615696649541a025d3d0f8e0447f-Paper.pdf.

Meng Liu, Tamal K. Dey, and David F. Gleich. Topological structure of complex predictions. *Nature Machine Intelligence*, Nov 2023a. ISSN 2522-5839. URL https://doi.org/10.1038/ s42256-023-00749-8.

Yajing Liu, Christina M. Cole, Chris Peterson, and Michael Kirby. ReLU neural networks, polyhedral decompositions, and persistent homology. In *2nd Annual TAG in Machine Learning*, 2023b.

Ziwei Liu, Ping Luo, Xiaogang Wang, and Xiaoou Tang. Deep learning face attributes in the wild. In *Proceedings of International Conference on Computer Vision (ICCV)*, December 2015.

Volker Lohweg. Banknote Authentication. UCI Machine Learning Repository, 2013. DOI: https:// doi.org/10.24432/C55P57.

Marissa Masden. Algorithmic determination of the combinatorial structure of the linear regions of ReLU neural networks, 2022. URL https://arxiv.org/abs/2207.07696.

Seyed-Mohsen Moosavi-Dezfooli, Alhussein Fawzi, Jonathan Uesato, and Pascal Frossard. Robustness via curvature regularization, and vice versa. In *Proceedings of the IEEE/CVF Conference on Computer Vision and Pattern Recognition (CVPR)*, June 2019.

Xiuyan Ni, Zhennan Yan, Tingting Wu, Jin Fan, and Chao Chen. A region-of-interest-reweight 3D convolutional neural network for the analytics of brain information processing. In Alejandro F. Frangi, Julia A. Schnabel, Christos Davatzikos, Carlos Alberola-López, and Gabor Fichtinger, editors, *Medical Image Computing and Computer Assisted Intervention – MICCAI 2018*, pages 302–310, Cham, 2018. Springer International Publishing. ISBN 978-3-030-00931-1.

Giovanni Petri and António Leitão. On the topological expressive power of neural networks. In *TDA & Beyond*, 2020. URL https://openreview.net/forum?id=I44kJPuvqPD.

Alec Radford, Luke Metz, and Soumith Chintala. Unsupervised representation learning with deep convolutional generative adversarial networks. In Yoshua Bengio and Yann LeCun, editors, *4th International Conference on Learning Representations, ICLR 2016, San Juan, Puerto Rico, May 2–4, 2016, Conference Track Proceedings*, 2016. URL http://arxiv.org/abs/1511.06434.

Karthikeyan Natesan Ramamurthy, Kush Varshney, and Krishnan Mody. Topological data analysis of decision boundaries with application to model selection. In Kamalika Chaudhuri and Ruslan Salakhutdinov, editors, *Proceedings of the 36th International Conference on Machine Learning*, volume 97 of *Proceedings of Machine Learning Research*, pages 5351–5360. PMLR, 09–15 Jun 2019. URL https://proceedings.mlr.press/v97/ramamurthy19a.html.

Michael W. Reimann, Max Nolte, Martina Scolamiero, Katharine Turner, Rodrigo Perin, Giuseppe Chindemi, Pawel Dlotko, Ran Levi, Kathryn Hess, and Henry Markram. Cliques of neurons bound into cavities provide a missing link between structure and function. *Frontiers in Computational Neuroscience*, 11, 2017. ISSN 1662-5188. URL https://www.frontiersin.org/articles/10.3389/fncom.2017.00048.

Mehdi S. M. Sajjadi, Olivier Bachem, Mario Lucic, Olivier Bousquet, and Sylvain Gelly. Assessing generative models via precision and recall. In S. Bengio, H. Wallach, H. Larochelle, K. Grauman, N. Cesa-Bianchi, and R. Garnett, editors, *Advances in Neural Information Processing Systems*, volume 31. Curran Associates, Inc., 2018. URL https://proceedings.neurips.cc/paper_files/paper/2018/file/f7696a9b362ac5a51c3dc8f098b73923-Paper.pdf.

Burr Settles. Active learning literature survey. Computer Sciences Technical Report 1648, University of Wisconsin-Madison, 2009. URL http://axon.cs.byu.edu/~martinez/classes/778/Papers/settles.activelearning.pdf.

Ilya Tolstikhin, Olivier Bousquet, Sylvain Gelly, and Bernhard Schoelkopf. Wasserstein auto-encoders. In *International Conference on Learning Representations*, 2018. URL https://openreview.net/forum?id=HkL7n1-0b.

Julius Von Rohrscheidt and Bastian Rieck. Topological singularity detection at multiple scales. In *International Conference on Machine Learning, ICML 2023, 23–29 July 2023, Honolulu, Hawaii, USA*, volume 202 of *Proceedings of Machine Learning Research*. PMLR, 2023.

Simon Willerton. Spread: A measure of the size of metric spaces. *International Journal of Computational Geometry & Applications*, 25(03):207–225, 2015. URL https://doi.org/10.1142/S0218195915500120.

Han Xiao, Kashif Rasul, and Roland Vollgraf. Fashion-MNIST: A novel image dataset for benchmarking machine learning algorithms, 2017. URL https://arxiv.org/abs/1708.07747.

Yuan Yuan, Eliezer M. Van Allen, Larsson Omberg, Nikhil Wagle, Ali Amin-Mansour, Artem Sokolov, Lauren A. Byers, Yanxun Xu, Kenneth R. Hess, Lixia Diao, Leng Han, Xuelin Huang, Michael S. Lawrence, John N. Weinstein, Josh M. Stuart, Gordon B. Mills, Levi A. Garraway, Adam A. Margolin, Gad Getz, and Han Liang. Assessing the clinical utility of cancer genomic and proteomic data across tumor types. *Nat Biotechnol*, 32(7):644–652, June 2014.

Sharon Zhou, Eric Zelikman, Fred Lu, Andrew Y. Ng, Gunnar E. Carlsson, and Stefano Ermon. Evaluating the disentanglement of deep generative models through manifold topology. In *International Conference on Learning Representations*, 2021. URL https://openreview.net/forum?id=djwS0m4Ft_A.

Zhengtong Zhu and Zhiyi Chi. Recursive computation of path homology for stratified digraphs, 2024. URL https://arxiv.org/abs/2412.05684.

Chapter 6
Internal Representations and Activations

Abstract Weights are arguably one of the most important parts of neural networks, as they determine their functions alongside architectures. Therefore, selecting the appropriate weights when training is key to ensuring that neural networks perform effectively. On the other hand, activations are the result of computations performed by a neural network on input data, and therefore are essential to understand the behavior of neural networks and their properties. The use of TDA to analyze weights and activations is divided into two approaches. The first one, more qualitative, uses Mapper to understand the structure of internal representations of neural networks. The second, more quantitative, uses persistent homology to extract topological features of internal representations of neural networks and link them with different properties of the networks, such as their generalization capabilities.

6.1 Mapper

This section is organized into two subsections: first, those articles that apply Mapper to point clouds derived from weights, and second, those that utilize Mapper on point clouds resulting from activations.

6.1.1 Mapper on Weights

Two seminal papers by Carlsson and Brüel Gabrielsson published in 2019 and 2020 stand as early explorations of the use of Mapper to analyze internal representations of FCFNNs, specifically CNNs. These papers focus on the topology and distribution of convolutional filter weights within the same layer of CNNs.

Consider a CNN with architecture a. Let i be a convolutional layer containing filters $C^i_{k,j}$ for $k, j \in [c^{(i-1)}] \times [c^{(i)}]$, each of dimensions $k_h^{(i)} \times k_w^{(i)}$. Here, $c^{(i-1)}$ denotes the number of channels in layer $i - 1$, and $c^{(i)}$ denotes the number of channels in layer i after convolution. Given t independently trained instances of this architecture, Carlsson and Brüel Gabrielsson (2020) studied Mapper graphs

© The Author(s), under exclusive license to Springer Nature Switzerland AG 2026

R. Ballester et al., *Topological Data Analysis for Neural Networks*, SpringerBriefs in Computer Science, https://doi.org/10.1007/978-3-032-08283-1_6

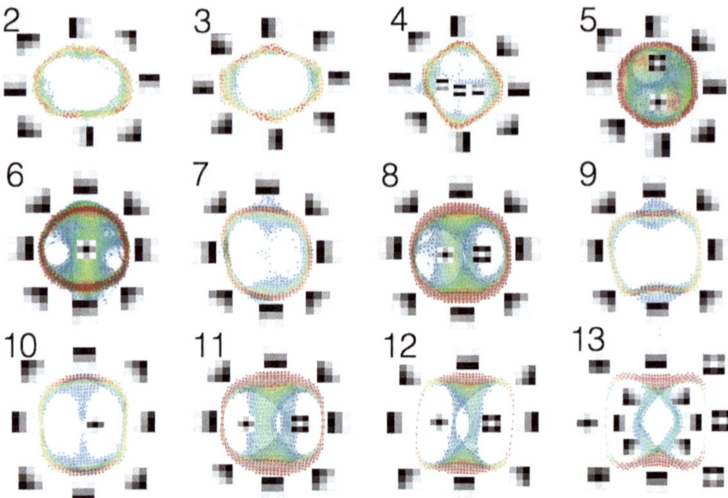

Fig. 6.1 Mapper graphs of the convolutional filter weights from layers 2 to 13 of a trained VGG-16 network. Vertex colors indicate size, increasing from blue to red. The matrices surrounding each Mapper graph are averages of filter weights located at nodes near the matrix. Image reproduced from Figure 10 in Brüel Gabrielsson and Carlsson (2019)

generated from a subset of the $tc^{(i-1)}c^{(i)}$ vectors of dimension $k_h^{(i)} \times k_w^{(i)}$ derived from the convolutions across the $c^{(i-1)}c^{(i)}$ filters in the t trained instances of the architecture.

In simple CNN architectures, comprising two convolutions, two pooling layers, and a fully connected output layer trained on the MNIST dataset, the Mapper graphs of the first convolutional layer closely resembled topological models of 3×3 patches of natural images that were previously identified by Carlsson et al. (2008). This suggests that CNNs may capture inherent topological features of data during learning.

Similar experiments were carried out on the CIFAR-10 dataset using architectures similar to those used for MNIST. Here, the results varied: some Mapper graphs resembled previously discovered topological models, while others exhibited new Mapper configurations. Mapper graphs of the more advanced architectures VGG-16 and VGG-19, pretrained on ImageNet, were also analyzed. For example, in VGG-16, all convolutional layers except the first and thirteenth ones displayed a circle as the dominant topological structure, consistent with findings in natural image patches. Also, it was observed that in the first layers of the network, the topological structures were simpler than in the other ones. Figure 6.1 displays the Mapper graphs for all convolutional layers in VGG-16, except the first one.

These Mapper graphs were also related to generalization. On the one hand, more powerful networks appeared to learn simpler topological structures in the first layers, as seen in VGG-16, suggesting a possible correlation between the topological complexity of the network's weights and its generalization capabilities. On the other hand, constraints on the parameter space of neural networks based on previously identified weight topological structures improved network generalization. Topological structures also helped preprocess the input data to obtain an increase in accuracy without modifying the network.

Building on such insights and the success of convolutional neural networks, Carlsson and Brüel Gabrielsson (2020) proposed three ways to generalize convolutions to layers that take into account the topology of data. This was continued in Love et al. (2020), where topological CNNs, which encompass several topologically defined convolutional methods, were introduced.

Gabella (2021) proposed to use Mapper to study the evolution of weights through the training process for a fixed layer of a FCFNN. Two different filter functions were used depending on the experiments: the ℓ^2 norm and projection onto the three principal components of the Mapper input points. Given a FCFNN trained in n steps and a layer l with N_l neurons, Mapper simplicial complexes were computed with a point cloud of $n \cdot N_l$ points of dimension N_{l-1}, each point representing the weights of the N_{l-1} incident edges to a neuron of layer l at a step of the training process.

In a first experiment, Gabella (2021) trained an FCFNN in MNIST with only one hidden layer of 100 neurons, setting the bias values to zero and initializing the weights to zero. The Mapper graph produced with an ℓ^2 filter function and the output layer had ten different branches, the same number as the number of classes in the dataset. For the hidden layer, only twelve distinct branches were produced, much less than the number of neurons. This suggests that the number of branches in this Mapper graph configuration could capture the *effective capacity* of a neural network layer —a concept explained informally as the layer's ability to learn and represent information from the input dataset. The branching patterns in the Mapper graph were correlated with the accuracy of the model in each training phase, supporting the claim.

In a second experiment, performed with an FCFNN with two hidden layers, each containing 100 neurons, and weights initialized randomly following a normal distribution, the Mapper graph for the output layer also contained ten branches. However, for the first hidden layer, PCA projections of the weights studied seemed to distribute in a tree evolving in parallel on top of a smooth surface, and a Mapper surface generated using the first three principal components of the points as filter function further confirmed this fact. This surface was remarkably robust under variations in the training parameters and the initial weights.

6.1.2 Mapper on Activations

Mapper has also been used to analyze the structure of neural network activations. TopoAct, by Rathore et al. (2021), was proposed as a visual exploration system to study the topology of activation vectors in fixed layers of neural networks. TopoAct uses the ℓ^2 norms of the activation vectors and DBSCAN as the filter function and the clustering algorithm, respectively. The cover of the image of the filter function is built using an algorithm considering a fixed number of covering sets with a fixed percentage of overlap between them. Finally, each Mapper node is augmented with some feature visualization techniques or statistical data to obtain more insight into the activations that comprise it.

One of the visualization techniques complementing the Mapper graphs proposed by Rathore et al. (2021) was the computation of *activation images* for the activations collected in each node and for their average. Activation images, given an activation vector or an average of them, are *idealized* images that would have produced the activations via an iterative optimization process proposed by Olah et al. (2017, 2018). Other statistical data proposed by Purvine et al. (2023) to describe Mapper nodes were the following:

1. Pie charts showing the composition of class labels in that node, associating to each activation vector the label of the image that generated the activation vector;
2. Node-wise purity, defined as $\alpha_i = c_i^{-1}$ where c_i is the number of class labels in the Mapper node i;
3. Class-wise purity, defined as $\gamma_y = |J_y|^{-1} \sum_{x \in J_y} \beta_x$, where J_y is the set of activations in the Mapper graph associated to the class y and $\beta_x = |I_x|^{-1} \sum_{i \in I_x} \alpha_i$, with I_x being the set of nodes in the Mapper graph containing the activation x.

The construction method for TopoAct input activation vectors varied depending on the type of layer under consideration. Convolutional layers emerged as the most extensively studied layers using TopoAct among the articles analyzed in this survey. For convolutional layers with output values of dimension $h \times w \times c$, Rathore et al. (2021) proposed to compute activations for each input value by randomly selecting two indices $i, j \in [h] \times [w]$ and taking the c dimensional vector obtained by fixing the dimensions i, j of the output. Later, two other sampling methods for activation vectors were proposed by Purvine et al. (2023):

1. Sampling all the possible $h \cdot w$ activations vectors for all pairs of indices i, j;
2. Sampling activations for which the indices i, j whose receptive fields are associated with the highest quantity of background or foreground pixels in the inputs are the ones selected.

For pretrained versions of Inception v1 (Szegedy et al. 2015), BERT (Devlin et al. 2019), and ResNet-18 (He et al. 2016), the Mapper graphs generated by TopoAct carried meaningful information. For Inception v1, Rathore et al. (2021) observed that branches in the Mapper graphs corresponded to activations containing different features of the inputs. For example, the activation images for one bifurcation showed that nodes of one branch were associated with animal legs, while nodes of the other were associated with distorted faces. Similarly, loops of nodes were associated with different aspects of the same underlying objects. An example was found in a loop that contained six nodes, each representing different features (body parts) of a set of animals, including dogs and foxes. For BERT, a language representation model, Mapper graphs showed a similar behavior separating word representations, where branches seemed to separate contextual meaning of similar words.

For ResNet-18, Rathore et al. (2021) observed branching patterns similar to those seen in the Inception v1 model, suggesting that TopoAct results are not specific to a particular dataset or architecture. Furthermore, Purvine et al. (2023) observed that the complete and random activation sampling methods for activation vectors yielded similar bifurcation patterns. For the sampling method corresponding to the highest number of background or foreground pixels in the input images, notable class bifurcations (that is, branches in the graph separating activations associated with different labels) seemed to appear at earlier layers, although this is not evident from the experiments. In general, node-wise and class-wise purity were found to be higher in deeper layers for all the sampling methods, confirming the idea that models get better at separating the classes the deeper they go.

Zhou et al. (2023) analyzed the impact of training neural networks with adversarial examples using TopoAct Mapper graphs derived from activations of deep layers of the networks on the training dataset. For this purpose, nodes were augmented using a variant of the node-wise purity introduced in Rathore et al. (2023). For a node i, the *purity* of i was defined as

$$1 - \frac{H(D_i)}{H(D)},$$

where H denotes the Shannon entropy of a distribution, D_i denotes the observed distribution of labels for the points in the node i, and D denotes a uniform distribution of all labels. This purity reaches 1 whenever all points in X are of the same class and 0 when points are distributed uniformly over all classes. The experiments were performed in two scenarios:

1. Training a simple FCFNN model with MNIST;
2. Training a ResNet-18 with CIFAR-10.

In the first scenario, where no overfitting occurred during training, the impure nodes —those without a dominant label— of the neural network trained without adversarial examples captured decision boundaries. Additionally, as the adversarial attack increased, i.e., the perturbations applied to the original examples grew larger, the number of nodes with low accuracy, meaning nodes with a high number of activations coming from misclassified inputs, in the Mapper graphs also increased.

In the second scenario, where overfitting was present, increased attack intensity led to a decrease in both the weighted average purity and the test accuracy. Here, the weighted average purity was calculated by weighting the purity of each node by the proportion of activations it contains relative to the total activations in the Mapper graph. The observation that the model overfits and that purity decreases with higher attack intensity suggests that, in this case, the impure nodes also captured the decision boundaries.

Based on such observations, Zhou et al. (2023) proposed to improve the robustness of adversarially trained neural networks by selecting misclassified activations from low-accuracy nodes on the Mapper graph and using them to refine the model. For the FCFNN network in MNIST, this procedure marginally improved the accuracy of the model. However, for the ResNet-18 in CIFAR-10, this procedure had no effect.

6.2 Homology and Persistent Homology

The literature has a wealth of methods to leverage the information provided by (persistent) homology features induced by activations and weights of neural networks. We divide this section into six blocks, each representing a space of internal representations in which TDA is applied, as follows:

1. Activations in the complete neural network graph;
2. Activations for each layer;
3. Weights in the complete neural network graph;
4. Weights layer by layer;
5. Activations whose dissimilarities depend on weights;
6. Generic spaces.

6.2.1 Activations in the Complete Neural Network Graph

Persistent homology of activation vectors has been successfully used to analyze many aspects of deep neural networks, such as their generalization or interpretability. In an early study within this section, Corneanu et al. (2019) analyzed how diverse topological information extracted from the set of neuron activations of neural networks was correlated with their generalization capabilities. The neuron activations were studied for the complete graph simultaneously, and the activation vector $a_{v_i^l}$ for each neuron v_i^l was taken as

$$a_{v_i^l} = \left(\phi_N^{(l)}(x_1)_i, \ldots, \phi_N^{(l)}(x_n)_i \right),$$

for a fixed set of inputs $\mathcal{D} = \{x_i\}_{i=1}^n$ to the neural network N, which is viewed as a subset of the training dataset.

More specifically, Corneanu et al. (2019) introduced *functional graphs*, with the objective of studying topological structures induced by correlations between neuron activations. The *functional graph* $\mathfrak{F}_N$ of a neural network N and subset of vertices P is the weighted, complete graph that includes a set P of neurons of N as its vertices, with edge weights given by the *correlation dissimilarity*

$$w_E(\{v, w\}) = 1 - \left| \mathrm{corr}(a_v, a_w) \right|,$$

where corr is the sample Pearson correlation and a_v and a_w are the activation vectors of neurons v and w, respectively. The graph $\mathfrak{F}_N$ can be treated as a point cloud

(P, d) where $d(v, w) = d(w, v) = w_E(\{v, w\})$ for $v \neq w$ and $d(v, v) = 0$ for all $v \in P$, making it suitable for TDA methods.

In their first experiments, Corneanu et al. (2019) studied how the number $S(n)$ of n-dimensional simplices of altered Vietoris–Rips simplicial complexes $\text{VR}_t^*(P, d)$, containing all the simplices of $\text{VR}_t(P, d)$ except for isolated vertices, varied during several training processes of a LeNet neural network architecture (Lecun et al. 1998) for a given parameter t satisfying

$$\frac{\left|\{\sigma = \{v, w\} \in \text{VR}_t^*(P, d)\}\right|}{|\{\sigma = \{v, w\} \subseteq P : |\text{corr}(a_v, a_w)| > 0\}|} \simeq 0.25.$$

In the experiments, it was found that, the higher the area under the curve of $S(n)$, the better the generalization capabilities of the neural network. In particular, it was shown that, for such training procedures, the function $S(n)$ is capable of distinguishing between the three main regimes given during training: underfitting, generalization, and overfitting. That is, the training process starts with a small area under $S(n)$ (underfitting regime), then $S(n)$ grows and achieves its maximum value (generalization regime), and then decreases again (overfitting regime).

These results motivated more involved experiments to find out the relationship between the topological properties of the activations of neural networks and their generalization capacity. For this purpose, Corneanu et al. (2019) computed persistence diagrams $D_k = D(\mathbb{V}_k(\text{VR}(P, d)))$ for $k \in \{1, 2, 3\}$ and a normalized version of their Betti curves

$$\beta_k(t) = \left| \{(b, d) \in D_k : b \leq t \leq d\} \right|,$$

during several training procedures of a LeNet neural network, with the expectation that persistent homology can provide more information than just counting the number of simplices.

During the first epochs, it was observed that the highest values of these curves moved from the left part of the domain to the right for $k \in \{1, 2, 3\}$. However, when the training entered into the overfitting regime, the maximum values of the normalized Betti curves moved again to the left part.

Due to the behaviour of the normalized Betti curves, Corneanu et al. (2019) proposed an early stopping of the training whenever it started moving the maximum values of Betti curves to the left again. Also, Corneanu et al. (2019) used normalized Betti curves to detect sets of adversarial examples, if any.

The observed evolution of Betti curves suggested that the topological structure of the functional graph of the activations of neural networks is highly related with their generalization capacity. This was further studied by Corneanu et al. (2020) and Ballester et al. (2024b), who used persistence diagrams $D(\mathbb{V}_k(\text{VR}(P, d)))$ in dimensions $k \in \{0, 1\}$ to predict the generalization gap, that is, the difference between training and test accuracies.

Corneanu et al. (2020) proposed to predict the generalization gap by performing a linear regression with independent variables chosen to be two persistence summaries, namely average persistence λ and average midlife μ, given by

$$\lambda(D) = \frac{1}{|D|} \sum_{(b,d) \in D} d - b, \qquad \mu(D) = \frac{1}{|D|} \sum_{(b,d) \in D} \frac{d + b}{2},$$

respectively. This approach seemed to work on controlled computer vision scenarios.

Following these results, Ballester et al. (2024b) extended previous work by studying more networks with different generalization gaps obtained from the first and second tasks of the *Predicting Generalization in Deep Learning* (PGDL) NeurIPS challenge (Jiang et al. 2020). In this study, additional persistence summaries were tested, such as persistent entropy (Atienza et al. 2020), persistence pooling vectors (Bonis et al. 2016), and complex polynomials (Di Fabio and Ferri 2015), among others.

The problem with the aforementioned approach is that it does not scale properly for the more complex networks analyzed in Ballester et al. (2024b). Due to the high number of neurons in larger neural networks and the high number of training examples in the datasets, computing persistence diagrams for the whole set of neurons and training examples was unfeasible. To alleviate this problem, Ballester et al. (2024b) proposed to generate multiple persistence diagrams from uniform samples of the dataset and from neuron samples using a probability distribution where neurons with higher activations in absolute value had a higher chance of being sampled than neurons with lower activations. From the corresponding persistence diagrams, persistence summaries were computed and then bootstrapped to perform linear regression to predict the generalization gap.

The best persistence summaries to predict generalization gaps in this line of work were a combination of statistical measures —such as averages and standard deviations— of the points in persistence diagrams. In particular, fixing some sources of variability of the networks affecting their depth, Ballester et al. (2024b) found a correlation between the averages and standard deviations of the second coordinates of zero- and one-dimensional persistence diagram points and the generalization gap of their associated networks. This correlation between topological summaries and generalization gap can be observed in Fig. 6.2. This figure shows that persistence summaries effectively detect clusters of neural networks for each task, aligning with the observation that the generalization gap is influenced by network depth. When the number of convolutions is fixed, a consistent behavior is observed: the higher the average deaths and the lower the standard deviations, the better the network's performance.

This discovery can be used for network regularization. A proof of concept was provided in a follow-up article by Ballester et al. (2024a), by developing TDA-based regularization terms that diminish the most significant correlations among neurons while still maintaining a degree of redundancy. By doing so, they outperformed classical regularization terms and improved performance compared to minimizing all pairwise correlations between neurons in the MNIST and CIFAR-10 datasets

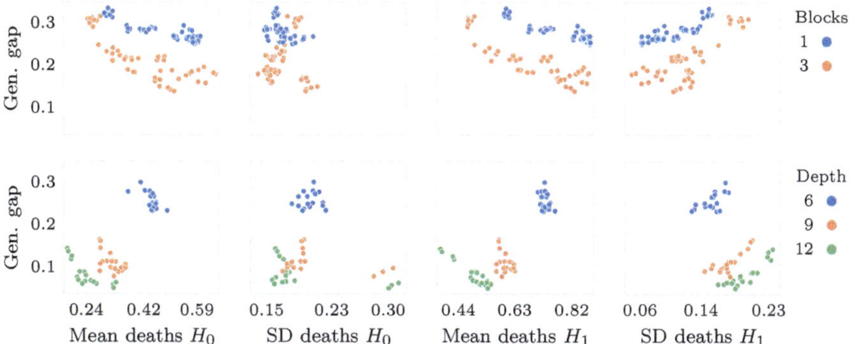

Fig. 6.2 Averages and standard deviations of deaths for persistence diagrams in dimension 0 (first two columns) and dimension 1 (last two columns) plotted for two groups of neural networks extracted from the PGDL challenge (Jiang et al. 2020) dataset against their generalization gap. The first row corresponds to the first group of neural networks, while the second row corresponds to the second group. For the first group, points represent 96 VGG-like neural networks trained on the CIFAR-10 dataset; blue and orange points represent neural networks with one and three convolutional blocks, respectively. For Task 2, points correspond to 54 network in network architectures trained on the SVHN dataset; blue, orange, and green points represent neural networks with six, nine, and twelve convolutional layers, respectively. Fixing the number of convolutional blocks or layers, we see a correlation between the generalization gap and the averages of the deaths, both in dimension zero and one. The higher the generalization gap, the lower the averages. This behaviour is shared for both sets of neural networks. Image reproduced from Figure 4.3 in Ballester et al. (2024b)

with multilayer perceptron and VGG-like architectures, respectively. This suggests that redundancies play a significant role in artificial neural networks, a phenomenon also observed in studies of human brain networks in neuroscience (Hennig et al. 2018).

A more specific study of one-dimensional persistence diagrams computed from neural network activations and their correlations was conducted by Zhang and Lin (2024). This study observed that the number of points in one-dimensional persistence diagrams and their 2-norms were correlated with distribution shifts in the input data and the network's training process. For image datasets, the number of points and the 2-norms of the persistence diagrams of the same network decreased as the dataset used to compute the activation vectors was increasingly corrupted by Gaussian blur or random pixel switching. Similarly, during training, the number of points and the 2-norms increased as the training process advanced until overfitting, where the number of points with high persistence stabilized suggesting an early stopping criterion without the need for a validation dataset, as in Corneanu et al. (2019).

A very similar approach was used to detect trojaned neural networks. A brief explanation of trojaned neural networks can be found in Sect. 4.4. Zheng et al. (2021) proposed to build zero- and one-dimensional Vietoris–Rips persistence modules from activations as in the previous works, yet replacing the Pearson correlation coefficient in absolute value by a more general correlation measure. For such persistence modules, Zheng et al. (2021) found a significant difference between clean and trojaned neural network persistence diagrams for both synthetic and real-world experiments, where the real-world experiments comprised training 70 ResNet18 using a clean MNIST dataset and another 70 with a trojaned MNIST counterpart.

The average death time of zero-dimensional persistence diagrams was particularly significant for segregating persistence diagrams from clean and trojaned neural networks. Averages for trojaned networks were significantly lower than those of their clean counterparts. Manual inspection also revealed that trojaned models exhibited edges connecting shallow and deep layers within cycle representatives of high-persistence points in persistence diagrams, a phenomenon absent in clean models.

With this in mind, Zheng et al. (2021) trained basic FCFNN models to detect trojaned convolutional neural networks on MNIST, CIFAR-10 and the IARPA/NIST TrojAI competition of Standards and Technology (2019) datasets using summaries from the persistence diagrams of the networks as input. These models resulted in higher performance than those for other state-of-the-art trojan detection algorithms.

6.2.2 Activations for Each Layer

While in previous works the topology of activations was studied globally, Naitzat et al. (2020) studied the evolution of activations layer by layer in FCFNNs for binary classification problems in an extensive set of experiments. For data samples $\mathcal{D}_i$ from one of the classes $i \in \{1, 2\}$ at a time, the evolution of the persistent homology of $\mathcal{D}_i$ through the different layers was analyzed. Activation vectors a_x^l given by

$$a_x^l = \left(\phi_N^{(l)}(x)_1, \ldots, \phi_N^{(l)}(x)_{N_l} \right)$$

were computed for each example $x \in \mathcal{D}_i$ and every layer l in an FCFNN N, representing the activations of each layer for the different input values. Then, persistent homology was calculated from Vietoris–Rips filtrations of the point clouds $\mathcal{D}_i^l = \left\{ a_x^l : x \in \mathcal{D}_i \right\}$ for each layer l.

The experiments were divided into two different scenarios: synthetic and real datasets. For the first scenario, the datasets were point samples from spaces with a known topology. For the second scenario, the datasets were simplified and binarized classification versions of several known datasets, such as MNIST. In both cases, FCFNNs were trained with different architectures, different training parameters, and almost zero training error.

For synthetic datasets, the evolution of Betti numbers was studied for different homological dimensions of Vietoris–Rips complexes of fixed value t. The dissimilarity used to build these complexes was chosen to be the graph geodesic distance on k-nearest neighbors graph applied to the point clouds $\mathcal{D}_i^l$. For a fixed class i and a fixed neural network $\mathcal{N}$, the parameters t and k were selected in such a way that $\beta_0(\mathrm{VR}_t(\mathcal{D}_i))$, $\beta_1(\mathrm{VR}_t(\mathcal{D}_i))$ and $\beta_2(\mathrm{VR}_t(\mathcal{D}_i))$ were equal to the first three (known) Betti numbers of the space from which the dataset $\mathcal{D}_i$ was sampled. For the second scenario, however, persistence diagrams of the activations for each layer were analyzed, circumventing the challenge of selecting optimal parameters t and k in real datasets.

In the synthetic scenario, Naitzat et al. (2020) observed a decay of the different Betti numbers across the layers. This decay was slower for zero-dimensional Betti numbers in networks implementing completely smooth activation functions than for networks implementing ReLU-like (ReLU and leaky ReLU) activation functions. The rate of decay correlated with the topological complexity of the input data. Specifically, datasets sampled from more intricate input spaces exhibited slower decay of Betti numbers compared with those from simpler spaces.

For neural networks with a constant number of neurons in the hidden layers, narrow layers —that is, those with few neurons per layer— appeared to simplify the topology faster than their wider counterparts. On the other hand, bottleneck architectures, i.e., architectures with decreasing numbers of neurons per layer, seemed to force larger topological changes in the data than networks with a constant number of neurons per layer.

Depth did not seem to influence the way the topological simplifications are distributed across the layers, since initial layers did not perform substantial topological simplifications, in general. Thus, reducing depth only had the effect of concentrating topological simplifications in the last layers. This topological simplification was also observed in the persistent diagrams of activations for the real datasets in simpler networks than the ones used for the synthetic scenario, where the number of points and their persistence across all dimensions diminished through layers.

In this work, Betti numbers were used to quantify the *topological complexity* of activations in a given layer. Persistence landscapes also allow one to define topological complexities over point clouds. Specifically, given a persistence landscape λ coming from a point cloud, one way to measure its complexity is to measure the area under its curves λ_i. The sum of these areas defines an inner product

$$\langle \lambda, \lambda' \rangle = \sum_{i=1}^{\infty} \int_{-\infty}^{\infty} \lambda_i(t)\lambda_i'(t)\, dt,$$

that induces a norm $\|\lambda\| = \sqrt{\langle \lambda, \lambda \rangle}$. The higher the norm, the higher the topological complexity of the point cloud.

Wheeler et al. (2021) used landscape complexities to contradict the previous results by Naitzat et al. (2020). In this work, evolution of the topology of activations through the different layers of FCFNNs for synthetic and real data was studied again. However, activations and distances between them were preprocessed and slightly modified depending on the experiments performed to build a point cloud. From such preprocessed point clouds, Vietoris–Rips persistence diagrams and their landscapes were computed using the Euclidean distance between vectors for each layer and in different homological degrees.

The results for synthetic data were similar to the previous ones: for FCFNNs with 11 layers with ReLU activation functions except for the output one (that used the identity function), trained to perfect accuracy with 100 different initialization weights, topological complexities were observed to decay through the last layers, in which the best weight configurations decayed faster on average than their worse counterparts. However, for the first layers, the topological complexities increased layer by layer. More strikingly, for FCFNNs trained to near-perfect accuracy with 7 layers distributed following a bottleneck architecture with ReLU activation functions except for the output one, topological complexities were observed to increase in the last layers on average and only decreased in the first layers.

Another promising approach, deeply related to the study of decision regions, is the analysis of the topology of activations of the output layer. The fundamental hypothesis is similar to what was found in Naitzat et al. (2020): the easier the topology of the last layer, the more robust, and thus the better, the neural network.

In this regard, Akai et al. (2021) studied the zero- and one-dimensional Vietoris–Rips persistence diagrams of last-layer activations, taken as in the previous articles, using the Euclidean distance. Although they did not prove the main hypothesis and could not relate network performance with output layer topology, it was a first step towards this direction.

Intuitively, the topology of activations for inputs of the same class must be similar. Based on this hypothesis, Zhao and Zhang (2022) proposed a way to measure the *quality* of convolutional filters for individual channels of convolutional layers in neural networks for classification problems. In this work, the inputs are squared images, and one assumes that the widths and heights of convolutional layer outputs are also equal. To compare the topology of the activations produced by the convolutional filters of interest, undirected weighted fully connected graphs are built. For a layer l, channel $c \in [c^{(l)}]$, and input sample x, the filter graph $C_x^{l,c}$ is generated so that $V(C_x^{l,c}) = [h^{(l)}]$ and

$$w_E(\{i, j\}) = \max\left(\phi_N^{(l)}(x)_{i,j,c}, \; \phi_N^{(l)}(x)_{j,i,c}\right),$$

for $i, j \in [h^{(l)}]$, where $\phi_N^{(l)}(x)_{i,j,c}$ is the (i, j, c) output value of N for the l-th convolutional layer given x.

Instead of directly comparing persistence diagrams from the previous graphs for the different data classes, Zhao and Zhang (2022) took a probabilistic approach based on the distribution of a specific topological summary for each label. Specifically, they computed for each graph the infimum of the support of the associated Betti curves coming from the persistence module $\mathbb{V}_1(\mathrm{VR}^1(V(C_x^{l,c}), d_\downarrow^0))$, which is directly related to the time in which the first *non-trivial cycle*, i.e., not given by a combination of triangles, appears in the filtration. We denote this minimum by $b_1^{\mathrm{inf}}(x, l, c)$.

The values $b_1^{\mathrm{inf}}(x, l, c)$ induce discrete random variables $B_1(l, c, y)$ for each label y of the classification problem with sample space Ω_y, i.e., the subset of samples from the training dataset $\mathcal{D}_{\mathrm{train}}$ with common label y. These random variables have probability distributions

$$P_{1,l,c,y}(n) = \frac{b_n}{\left|\Omega_y\right|}, \qquad b_n = \sum_{x \in \Omega_y} \mathbb{1}_{\left\{x' : b_1^{\mathrm{inf}}(x',l,c)=n\right\}}(x),$$

that capture information about the similarity of the topology of the activations given by the convolutional filters of interest in the data distributions of the different labels of the classification problem. This similarity can be further quantified using the entropies of the probability distributions, given by

$$H'_{1,l,c,y} = -\sum_{n=0}^{\infty} P_{1,l,c,y}(n) \log P_{1,l,c,y}(n). \tag{6.1}$$

The values given by (6.1) are called *feature entropies*, and are the topological summaries proposed by Zhao and Zhang (2022) to measure the quality of convolutional filters. Feature entropies are invariant to weight reescaling, a desirable property to measure the quality of the convolution operations, as reescalings of weights have no substantial impact on the network performance in general.

One problem of this approach is that the infimum value b_1^{inf} may not exist in some cases. For layers and channels in which this happens many times, the entropy approaches zero, although entropy is not really informative as it is affected by the non-existence of infimum values. In such cases, entropy is modified as

$$H_{1,l,c,y} = \begin{cases} H'_{1,l,c,y} & \text{if } \varepsilon_{1,l,c,y} \geq p, \\ (1 - \varepsilon_{1,l,c,y}) \log\left|\Omega_y\right| & \text{otherwise,} \end{cases}$$

where $\varepsilon_{1,l,c,y}$ is the percentage of images in class y having birth times, and p is the minimum percentage of images that we admit to use the real entropy (in the article, $p = 0.1$).

Zhao and Zhang (2022) demonstrated the effectivity of the entropy measure to obtain information on the *quality* of the convolutions and of the entire neural network for VGG-16 models trained on the ImageNet dataset. They observed that, for well-trained neural networks, feature entropy continually decreased as the layers went deeper, while for random weights this decreasement was absent. On the other hand, they observed that, during training, the feature entropies of the last convolutional layer decreased, and were highly correlated with the evolution of the cross entropy training loss, suggesting that feature entropies are good indicators of the generalization of networks. Additionally, randomness of the weights was also detected by comparing the feature entropies of trained and randomly initialized networks. Finally, for the last convolutional layers, models with better generalization were correlated with low feature entropies.

A similar study was conducted by Yang et al. (2024), who analyzed the evolution of Betti curves induced by persistence diagrams computed from convolutional filter weights during training. For a fixed set of convolutional filters $\{C_i\}_{i \in I}$ of the network with the same size $h \times w$ (not necessarily from the same layer), convolution filters were transformed into vectors of size $h \cdot w$ and then used as points in a point cloud $(P, \|\cdot\|_2)$ to compute persistence diagrams of dimensions zero and one, from which Betti curves for these dimensions were extracted.

The evolution of Betti curves during training was studied for two different convolutional architectures and several datasets, including grayscale images, color images, and noise images. The networks consisted of three or four convolutional layers, depending on the architecture, and two fully connected layers. Additionally, the sizes of the convolutional filters were fixed to 3×3 across all the layers. The subset of convolutional filters used to compute persistence diagrams included all filters except those from the first convolutional layer.

For the grayscale and color datasets, the Betti curves of zero- and one-dimensional persistence diagrams were observed to shift rightward as training iterations (and performance) increased. For one-dimensional Betti curves, which exhibited a bell-shape, the maximum value decreased as training progressed. For the noise datasets, however, Betti curves remained stable across iterations, suggesting that learning occurred when changes in Betti curves happened during training, as there are no patterns to learn in random noise. Finally, in a promising application of TDA to modern Large Language Model (LLM) architectures, Gardinazzi et al. (2024) used zigzag persistence to track topological changes in layer activations. This led to a novel, topologically informed pruning strategy that achieved performance comparable to current state-of-the-art methods.

6.2.3   *Weights in the Complete Neural Network Graph*

Recall that the input values influence the output of a neural network via the different input-output paths available in the neural network graph. The influence of each path is characterized by the magnitudes of the weights associated with the edges in the path, which modify the magnitudes of the activations, and thus their relevance, at each step. Watanabe and Yamana (2022b) proposed to build neural network graph filtrations taking into account the influence of the different paths in the network.

To formalize their approach, let $\mathcal{N}$ be a neural network, and $v_l^i, v_{l+1}^j \in V(G(\mathcal{N}))$ be two connected neurons in the directed graph $G(\mathcal{N})$. We define the *relevance* of the directed edge (v_l^i, v_{l+1}^j) connecting v_l^i and v_{l+1}^j in the output calculation as the linearly rectified weight of the edge, normalized by the sum of the linearly rectified weights of edges pointing to v_{l+1}^j, that is,

$$R_{v_l^i, v_{l+1}^j} = \frac{\mathrm{ReLU}\big(W_{i,j}^{(l+1)}\big)}{\sum_{k \neq i} \mathrm{ReLU}\big(W_{k,j}^{(l+1)}\big)}.$$

This edge-level relevance can be extended to define relevance between vertices connected by a path in $G(\mathcal{N})$. For any two vertices v and w connected by a path, the influence from v to w is characterized as the maximum product of relevances of edges among all paths between v and w, whenever $v \neq w$, and 1 otherwise:

$$\tilde{R}_{v,w} = \begin{cases} \max_{(v, p_1, \ldots, p_n, w) \in P_{v,w}} R_{v, p_0} \left(\prod_{i=1}^{n-1} R_{p_i, p_{i+1}} \right) R_{p_n, w} & \text{if } v \neq w, \\ 1 & \text{otherwise,} \end{cases}$$

where $P_{v,w}$ is the set of all paths starting at v and ending at w in $G(\mathcal{N})$.

Building upon these definitions, the proposed graph filtrations were defined as filtrations $(K_i)_{i=1}^n$ with a monotonically decreasing sequence of indices $(t_i)_{i=1}^n$ given by

$$K_i^p = K_{t_i}^p = \begin{cases} V(G(\mathcal{N})) & \text{if } p = 0, \\ \{\{v_{k_0}, \ldots, v_{k_p}\} : \tilde{R}_{v_{k_i}, v_{k_j}} \geq t_i \text{ for } k_i > k_j\} & \text{if } p \geq 1, \end{cases}$$

where p indicates the dimensions of the simplices of K_i^p, and the vertices $V(G(\mathcal{N}))$ are ordered in such a way that if the layer number of v_i is lower or equal than the layer number of v_j then $i \geq j$.

The filtration $(K_i)_{i=1}^n$ captures the influence of the different paths between neurons in the neural network graph and can be easily extended to a persistence module where the simplicial complex assigned to time $t \in \mathbb{R}$ is $K_{\lfloor t \rfloor}$ for $0 \leq t \leq n$, K_0 for $t < 0$, and K_n for $t > n$.

For simple networks trained on MNIST and CIFAR-10, the distribution of points of one-dimensional persistence diagrams computed from persistence modules coming from the previous filtration appeared to be correlated with several properties of the network and the dataset, such as problem difficulty or network expressivity. For example, the appearance of points near the diagonal of the persistence diagrams was associated with a shortage of data of specific labels in the training dataset.

In addition, persistence diagrams seemed to be robust with respect to the initial values of the neural network weights, yielding each architecture similar persistence diagrams for different initializations after training. However, a more thorough study of such persistence modules and their associated persistence diagrams is needed to understand their relationship with the generalization capabilities of a neural network, as Watanabe and Yamana (2022b) remarked.

In a follow-up work, Watanabe and Yamana (2022a) presented an extension of the relevance quantity for convolution and pooling transformations that led to persistence modules better suited to analyze convolutional neural networks. Specifically, Watanabe and Yamana (2022a) used this extension to study the overfitting of CNNs. They observed that, in simple scenarios, distributions of points in one-dimensional persistence diagrams were correlated with the overfitting of convolutional networks. Specifically, as the dropout rate increased for different training instances within the same architecture, a corresponding increase in the number of points near the diagonal and in overfitting was also observed. Furthermore, it was found that a higher total number of points in the persistence diagram correlated with reduced underfitting.

One drawback of their method is that straightforward persistence diagrams were found not to be entirely suitable for comparing models with different architectures in this case. To address this difficulty, Watanabe and Yamana (2022a) proposed a novel approach: pruning the networks of interest using a magnitude-based strategy (Blalock et al. 2020) before generating persistence diagrams to compare them so that the persistence diagrams were normalized with respect to the size of the network. After selecting several groups of neural networks to be compared, an overall high correlation between the number of points near the diagonal of normalized persistence diagrams and overfitting of the networks of the different groups was observed, validating their previous results in a more general setting.

Pruning using persistent homology was further developed in Watanabe and Yamana (2020) using a slightly modified relevance function given by

$$R_{v,w} = \frac{\left|W_{i,j}^{(l+1)}\right|}{\sum_{k \neq i}\left|W_{k,j}^{(l+1)}\right|}.$$

Watanabe and Yamana (2020) proposed a pruning algorithm that was shown to be competitive with respect to the global magnitude algorithm (Blalock et al. 2020) in terms of final accuracy of the pruned networks. The algorithm was as follows:

1. Sort points (b, d) of one-dimensional persistence diagrams by their value $b + d$ in ascending order;
2. For each point, choose a cycle representative c of the point (b, d);
3. Select the edges contained in the representatives in the previous step until you reach a desired number of edges to retain;
4. Prune the edges not selected in the previous step.

A similar approach to the one taken by Corneanu et al. (2020) and by Ballester et al. (2024b) was proposed by Barbara et al. (2024), who argued that the dataset has too much influence on the activations and thus their correlations do not capture all the relevant information about the neural network related with generalization. Instead, they proposed to analyze zeroth persistence diagrams $D(V_0(\mathrm{VR}(P, d)))$ induced by subsets P of neural network vertices and dissimilarities given by the Euclidean distance between HVS scores S_v (Yacoub and Bennani 1997) associated to each vertex v of P, defined as

$$S_v = \sum_{(v,w)\in E(G(\mathcal{N}))} \pi_{w,v} \cdot \delta_w,$$

$$\pi_{w,v} = \frac{\left|W_{w,v}\right|}{\sum_{(u,w)\in E(G(\mathcal{N}))} \left|W_{w,u}\right|}, \qquad \delta_w = \begin{cases} 1 & \text{if } w \in V_L, \\ S_w & \text{otherwise,} \end{cases}$$

where $W_{w,v}$ is the weight corresponding to the edge $(v, w) \in E(G(\mathcal{N}))$ and V_L is the set of neurons of the output layer.

To study the relationship of these diagrams with the generalization gap, Barbara et al. (2024) trained very simple FCFNNs with two hidden layers for a variety of datasets from the UCI Machine Learning repository (Kelly et al. 1987), taking P as the 90% of neurons with highest relevance score for each network. For the experiments, the generalization gap was compared with the average persistence value of the diagram generated at each iteration of the training procedure. Linear regression models were fitted with the generalization gaps and average persistences during training as dependent and independent variables, respectively. The R^2 values of these linear regressions were slightly greater than the ones for the same linear regression models using the persistence diagrams computed in Ballester et al. (2024a). However, the simplicity in the experiments performed, both in the experimental pipeline and in the networks used, makes that further experimentation is needed to evaluate if this approach generalizes to more complex scenarios and also improves the methods of Ballester et al. (2024a) in them.

6.2.4 Weights Layer by Layer

Layerwise topological complexities have also been studied as a function of the weights. Rieck et al. (2019) proposed to study the so-called neural persistence. Given a non-output layer l of a neural network $\mathcal{N}$, its *neural persistence* is defined as

$$\mathrm{NP}_l(\mathcal{N}) = \|D\|_p = \left(\sum_{(b,d)\in D} |d - b|^p \right)^{1/p},$$

where $D = D^w\big(\mathbb{V}_0\big(\mathrm{VR}^1\big(V(G_l), d_{\downarrow}^V\big)\big)\big)$ is the zeroth persistence diagram of the weighted complete bipartite graph G_l with vertices $V_l \sqcup V_{l+1}$ and weights given by $w_V(v) = w_{\max}$ and $w_E(\{v_l^i, v_{l+1}^j\}) = |W_{j,i}^{(l+1)}|/w_{\max}$, where $W_{j,i}^{(l+1)}$ is the weight associated to the edge connecting the vertices v_l^i and v_{l+1}^j, as in Fig. 2.1, and $w_{\max} = \max_{l,i,j} |W_{i,j}^l|$ is the maximum weight in absolute value among all the weights of the network.

To compare the neural persistence of different layers from the network, Rieck et al. (2019) proposed to normalize neural persistence by dividing it by an upper bound of the neural persistence,

$$\mathrm{NP}_l^+(\mathcal{N}) = w_{\max}^{-1}\left(\max_{i,j}\left|W_{i,j}^{(l+1)}\right| - \min_{i,j}\left|W_{i,j}^{(l+1)}\right|\right)(N_l - 1)^{1/p}\,,$$

obtaining the normalized neural persistence $\widetilde{\mathrm{NP}}_l(\mathcal{N})$ for each layer l. Averaging normalized neural persistences over all the layers of the network, a global topological complexity measure $\overline{\mathrm{NP}}(\mathcal{N})$ is obtained.

For simple FCFNN networks trained on MNIST, neural persistence was able to distinguish clearly between properly and badly trained networks, for which the values of the neural persistence of properly trained ones were consistently higher than for their badly trained counterparts. Furthermore, it was observed that different regularization techniques augmented the mean neural persistence with respect to the values of regular trained networks, suggesting that for a fixed architecture, the higher the neural persistence, the better the generalization capabilities of the network.

This was exploited as an early stopping criterion for training neural networks, where the training process is completed when the mean neural persistence $\overline{\mathrm{NP}}(\mathcal{N})$ stops increasing significantly. This early stopping criterion was found to be competitive with other early stopping criteria but without the need for a validation dataset.

Girrbach et al. (2023) deepened more into the properties of neural persistence. A key contribution of their work was observing a close relationship between the variance of the learned weights of deep learning models and the neural persistence, questioning the value of the latter with respect to this simpler measure to study the properties of the neural network, arguing that this variance may be similarly useful for the applications showcased by Rieck et al. (2019). Finally, some preliminary recent work (Gardinazzi et al. 2024) has proposed the use of TDA for the study of LLM internal representations.

6.2.5 *Activations Whose Dissimilarities Depend on Weights*

Both weights and activations can be used together to study the topology of neural networks. In particular, they can be used to inspect differences in the internal workings of neural networks for different input values. Given an input sample x and a neural network $\mathcal{N}$, Gebhart et al. (2019) proposed to analyze the zeroth persistence module $\mathbb{V}_0\big(\mathrm{VR}^1\big(V(G_x), d_\downarrow^0\big)\big)$ for G_x the undirected weighted graph induced by the directed neural network graph $G(\mathcal{N})$ with weights given by the formula

$$w_E\big(\{v_l^i, v_{l+1}^j\}\big) = \big|W_{j,i}^{(l+1)}\phi_{\mathcal{N}}^{(l)}(x)_i\big|.$$

The weight function captures how activations are distributed through the neural network, as in the previous works. However, in this case, activations are weighted by the weights of the neural network, which do not necessarily depend on the example x, and which have an effect of normalization on the influence of the input. Each persistence module is intended to capture information of the network under the influence of input x. In this way, the differences between persistence modules for the same network and different inputs can be used to gain insight into the dynamics and representations used by the neural network to compute its function.

To measure the differences between persistence modules coming from the same network $\mathcal{N}$ and different inputs x, x' quantitatively, Gebhart et al. (2019) proposed to calculate a dissimilarity based on the differences between the cycle representatives of the points of the two persistence diagrams $D_x = D\big(\mathbb{V}_0\big(\mathrm{VR}^1\big(V(G_x), d_\downarrow^0\big)\big)\big)$ and $D_{x'} = D\big(\mathbb{V}_0\big(\mathrm{VR}^1\big(V(G_{x'}), d_\downarrow^0\big)\big)\big)$, respectively.

More precisely, let $\{\alpha_i\}_{i=1}^{|D_x|}$ and $\{\beta_j\}_{j=1}^{|D_{x'}|}$ be representatives of the points of D_x and $D_{x'}$, respectively. Each representative cycle α_i and β_j is associated with a subgraph of the graph G_x and $G_{x'}$, respectively, denoted by T_i and T_j'. Let v_x and v_x' be two vectors with as many components as edges there are in the union of graphs

$$\left(\cup_{i=1}^{|D_x|} T_i\right) \cup \left(\cup_{j=1}^{|D_{x'}|} T_j'\right),$$

respectively, where v_x and v_y have one in the components corresponding to the edges that come from the union of graphs $\left(\cup_{i=1}^{|D_x|} T_i\right)$ and $\left(\cup_{j=1}^{|D_{x'}|} T_j'\right)$, respectively, and a zero otherwise. Then, the dissimilarity between the persistence modules generated from x and x' is defined as a weighted version of the Hamming distance between v_x and $v_{x'}$ using the persistence of the cycle representatives associated with the edges corresponding to the different components of the vectors as weights.

Gebhart et al. (2019) demonstrated the efficacy of the aforementioned persistence modules and associated dissimilarity measures in analyzing neural networks through experiments on MNIST, FashionMNIST, and CIFAR-10 datasets, employing three simple convolutional architectures, including an AlexNet (Krizhevsky et al. 2012) variant for the CIFAR-10 dataset. Their findings were threefold:

1. Persistence modules effectively captured distinctions between clean and adversarial or shifted examples when processed by the networks;
2. Distances between persistence modules derived from image sets closely mirrored actual image distances;
3. Support vector machines (SVMs) achieved high classification accuracy using persistence modules as input, with the dissimilarity measure serving as a kernel.

These results suggest that the topological information extracted by persistence modules encapsulates much of the data used by neural networks for classification. Moreover, this topological information appears sufficiently distinct to elucidate differences in the neural network's behavior across various classes.

The aforementioned persistence modules were further studied to detect adversarial examples by Goibert et al. (2022). Their hypothesis was that only a small set of edges within the networks are used for inference of non-adversarial inputs, meaning that activations associated to most edges are irrelevant to the output, and that, for adversarial examples, the number of edges used for inference is larger. The idea behind the hypothesis is that adversarial examples attack input-output edge paths containing underused edges of the neural network to, with imperceptible modifications of the inputs, completely change the output.

Ideally, changes in the activations of neurons within underused paths lead to structural modifications in the activation patterns of the neural networks, consequently affecting their persistence modules. To better identify topological changes, Goibert et al. (2022) only added edges to the graph G_x that were underoptimized, defined as those with *low* weights in absolute value. Their hypothesis was reflected in the number of points of zeroth persistence diagrams points coming from persistence modules of adversarial and non-adversarial examples, where adversarial examples had more points on average than their non-adversarial counterparts. Furthermore, they successfully detected adversarial examples by training a support vector machine using the sliced Wasserstein kernel (Carrière et al. 2017) between persistence diagrams for a variety of neural network types (LeNet, ResNet), datasets (MNIST, Fashion-MNIST, SVHN, CIFAR-10) and adversarial attacks.

Topological data analysis has also been used to analyze neural networks in reinforcement learning (RL) tasks. Muller et al. (2024) studied the evolution of Betti numbers given by homology groups of complexes induced by graphs coming from RL neural networks during a *time* period either in inference or training. Specifically, homology groups for the directed flag (Lütgehetmann et al. 2020) and Vietoris–Rips complexes were calculated from adjacency graphs $G_x^{r,d}$ derived from the weighted activation graph G_x proposed by Gebhart et al. (2019), where x was the input of the RL agent at a specific time. Each graph $G_x^{r,d}$ was built by taking the edges of G_x with weights higher than or equal to r and adding an edge between two vertices v_l^i and v_l^j if there existed a third vertex v_{l-1}^k such that

$$\left| \left(W_{i,k}^{(l)} - W_{j,k}^{(l)} \right) \phi_N^{(l-1)}(x)_k \right| < d,$$

meaning that the activations of vertices v_l^i and v_l^j are highly correlated for input x.

For inference experiments, Betti numbers up to dimension three for each time step were computed and smoothed using a moving average window. Transitions between actions of the agent and evolution of the Betti numbers of dimension three seemed to be correlated, while for the other dimensions this correlation was not conclusive. For training experiments, complex environments seemed to induce the development of higher-order Betti numbers during training. Also, with a similar number of neurons, the higher the Betti numbers in the last steps of training, the better the model worked.

Training was also studied through a matrix H with training steps as columns and neurons as rows. Each coefficient $H_{v,t}$ of H was derived from cocycle representatives of one-dimensional cohomology groups of Vietoris–Rips complexes at time t. Specifically, $H_{v,t}$ was the maximum length of 1-cocycles in the set C_v^t of cocycles whose support had edges containing the vertex v at time t.

The matrix H was seen to have higher values for neurons on the latest layers consistently for all the training steps, suggesting that deeper neurons have more relevance in the topological structure extracted from the graphs $G_x^{r,d}$. However, more experiments are needed to validate these results, as experiments were performed only for basic FCFNNs.

A combination of the methods proposed in Rieck et al. (2019) and in Gebhart et al. (2019) was used to define a new prediction reliability score for neural networks in classification problems called *topological uncertainty*. Given a neural network $\mathcal{N}$ and a sample x_{in} for which we want to compute this score, the topological uncertainty of x_{in} is computed from the set of persistence diagrams

$$D_x^l = D^w\big(\mathbb{V}_0\big(\text{VR}^1\big(V(G_x^l), d_V^0\big)\big)\big),$$

for all $l \in [L]$ and $x \in \{x_{\text{in}}\} \cup \{x : (x, y) \in \mathcal{D}_{\text{train}}\}$, where L is the number of non-input layers of $\mathcal{N}$, and $\mathcal{D}_{\text{train}}$ is the training dataset, and G_x^l is the subgraph of G_x as in Gebhart et al. (2019) induced by the vertices of the layer l and $l-1$ of $\mathcal{N}$ and all the possible edges connecting them in G_x.

In this case, edges are weighted as in G_x but vertices are weighted as -1. This is done to always have a fixed number of points in zeroth persistence diagrams equal to the number of vertices of G_x^l for any possible weight assignment to the edges and to have all birth values equal. It can be proved that, in this way, there is a bijection between the finite deaths of the persistence diagrams D_x^l and the multiset of weights of the maximum spanning tree of G_x^l, which we denote by $\mathfrak{W}_x^l$. These persistence diagrams are meant to capture information similar to that computed in Gebhart et al. (2019), but more fine-grained, since they are computed for each pair of adjacent layers.

Maximum spanning tree weights are used to build class-specific weight distributions. The topological uncertainty measure for an example is then computed as the difference between its maximum spanning tree weights and the distribution of weights for its predicted class. Recall that $\left|\mathfrak{W}_x^l\right| = \left|\mathfrak{W}_{x'}^l\right|$ for any pair of inputs x and x'. Given a multiset of weights $\mathfrak{W}_x^l$, let us order them in descending order and denote them by

$$\mathfrak{W}_x^l = \left\{w_{x,1}^l, \ldots, w_{x,\left|\mathfrak{W}_x^l\right|}^l\right\}.$$

Weights induce a probability distribution for each subgraph G_x^l given by

$$\mu_x^l = \frac{1}{n}\sum_{i=1}^n \delta_{w_{x,i}^l},$$

where $n = \left|\mathfrak{W}_x^l\right|$ and $\delta_{w_{x,i}^l}$ denotes a Dirac measure at $w_{x,i}^l \in \mathbb{R}$. For the same layer, these probability distributions can be combined to obtain an *average topological distribution* across several samples. The average over points $x_1, \ldots, x_m$ is given by

$$\bar{\mu}^l = \frac{1}{\left|\mathfrak{W}_x^l\right|} \sum_{i=1}^{\left|\mathfrak{W}_x^l\right|} \delta_{\bar{w}_i}, \qquad \bar{w}_i = \frac{1}{m} \sum_{j=1}^{m} w_{x_j, i}^l.$$

Therefore, given the new sample x_{in}, the topological uncertainty measures the average difference between the average topological distributions $\bar{\mu}^l$ for each layer l calculated for the subset of the training dataset $\mathcal{D}_{\text{train}}$ with label equal to the predicted label for x_{in} with respect to the topological distributions $\mu_{x_{\text{in}}}^l$ of the input sample x_{in}, that is,

$$\text{TU}_x(\mathcal{N}) = \frac{1}{L} \sum_{l=1}^{L} \text{d}\big(\mu_{x_{\text{in}}}^l, \bar{\mu}_{\mathcal{D}_x}^l\big),$$

$$\mathcal{D}_x = \{x : (x, y) \in \mathcal{D}_{\text{train}} \text{ with } \pi \circ \phi_{\mathcal{N}}(x_{\text{in}}) = y\},$$

where $\bar{\mu}_{\mathcal{D}_x}^l$ is the average probability distribution over the set $\mathcal{D}_x$ and where

$$\text{d}(\mu_1, \mu_2) = \frac{1}{\left|\mathfrak{W}_1\right|} \sum_{i=1}^{\left|\mathfrak{W}_1\right|} \left| w_i^1 - w_i^2 \right|$$

is a dissimilarity function between distributions coming from sets of weights $\mathfrak{W}_1 = \left\{w_1^1, \ldots, w_{|\mathfrak{W}_1|}^1\right\}$ and $\mathfrak{W}_2 = \left\{w_1^2, \ldots, w_{|\mathfrak{W}_1|}^2\right\}$ with the same number of points decreasingly ordered as before.

The lower the value of the topological uncertainty measure, the more reliable the prediction for x_{in}, since its internal behavior is more similar to the behaviour of the network for the examples in the training dataset with the same label. This measure was successfully used for model selection from a set of models trained on MNIST and Fashion-MNIST, where the model with the lowest average topological uncertainty for the new dataset is selected. It was also used for detecting out-of-distribution and shifted examples in several basic networks and datasets, including MUTAG (Debnath et al. 1991), COX2, and MNIST datasets, considering only some set of (non-convolutional) layers to compute topological uncertainty.

6.2.6   *Generic Spaces*

So far we have seen methods to compare differences between the topology, in a broad sense, of different neural network *representations*. Most of these methods are based on distances on either persistence diagrams or on some constructions coming from persistence modules. A more direct approach is taken by Barannikov et al. (2022), who propose a method to compare two Vietoris–Rips filtrations for the same set of points V and different distances d_1, d_2. The idea behind the method is to compare, given a specific threshold t, how the connected components of $\mathrm{VR}_t(V, d_1)$ and $\mathrm{VR}_t(V, d_2)$ are merged in $\mathrm{VR}_t(V, d_{\min})$, where $d_{\min}(x, y) = \min(d_1(x, y), d_2(x, y))$. Hence, they count how many connected components are merged at each threshold for all possible thresholds, and derive a measure of dissimilarity from such countings.

This measure of dissimilarity is called *representation topology divergence* and is defined as the average of two total persistences, denoted by $\mathrm{RTD}_1(d_1, d_2)$ and $\mathrm{RTD}_1(d_2, d_1)$, calculated from one-dimensional persistence diagrams of the Vietoris–Rips filtrations of the point clouds $(V_{1,2}, d_{1,2})$ and $(V_{2,1}, d_{2,1})$ with vertices $V_{i,j} = \{v_a\}_{a=1}^{|V|} \cup \{v_a'\}_{a=1}^{|V|} \cup \{O\}$ and distances

$$d_{i,j}(v_a', v_b') = \min(d_i(v_a, v_b), d_j(v_a, v_b)), \qquad d_{i,j}(v_a, v_a') = d_{i,j}(O, v_a) = 0,$$

$$d_{i,j}(v_a, v_b') = d_{i,j}(v_a, v_b) = d_i(v_a, v_b), \qquad d_{i,j}(v_b, v_a') = d_{i,j}(O, v_a') = \infty,$$

for $i \in [2]$, $j \in [2] \setminus \{i\}$, where v_a' is the node v_a duplicated in the point cloud and O is an abstract point that is useful to capture the differences between $\mathrm{VR}_t(V, d_i)$, $\mathrm{VR}_t(V, d_j)$ and $\mathrm{VR}_t(V, d_{\min})$.

Intuitively, the k-dimensional persistence diagram from the Vietoris–Rips filtration of these point clouds records the k-dimensional topological features that are born in $\mathrm{VR}_t(V, d_{\min})$ but not yet in $\mathrm{VR}_t(V, d_i)$, and the $(k - 1)$-dimensional topological features that are dead in $\mathrm{VR}_t(V, d_{\min})$ but are not yet dead for $\mathrm{VR}_t(V, d_i)$.

The representation topology divergence has been used to analyze many aspects of neural networks. For example, Barannikov et al. (2022) used it to analyze 400-dimensional embeddings of 10,000 words for 90 randomly selected architectures from the NAS-Bench-NLP (Klyuchnikov et al. 2020), as well as differences between real and shifted data, and several properties of neural networks, among others. For studying neural network properties, they trained a VGG-11 and a ResNet-20 convolutional networks on CIFAR-10 and CIFAR-100 and compared the evolution of the activations of convolutional layers to the activations of the final trained network for the same layer.

They observed that the representation topology divergence between activations at each epoch during training and activations of the network after training decreased as the number of epochs trained increased, capturing the convergence to the final state of the network. They also found that the evolution of the representation topology divergence for both models correlated well with the disagreement of the predictions between them. Finally, they showed that the representation topology divergence could be a good indicator of diversity and disentanglement of models in an ensemble method or generative tasks, respectively, when used properly.

References

Naoki Akai, Takatsugu Hirayama, and Hiroshi Murase. Experimental stability analysis of neural networks in classification problems with confidence sets for persistence diagrams. *Neural Networks*, 143:42–51, 2021. ISSN 0893-6080. https://doi.org/10.1016/j.neunet.2021.05.007. URL https://www.sciencedirect.com/science/article/pii/S0893608021001994.

Nieves Atienza, Rocio Gonzalez-Diaz, and Manuel Soriano-Trigueros. On the stability of persistent entropy and new summary functions for topological data analysis. *Pattern Recognition*, 107:107509, 2020. ISSN 0031-3203. https://doi.org/10.1016/j.patcog.2020.107509. URL https://www.sciencedirect.com/science/article/pii/S0031320320303125.

Rubén Ballester, Carles Casacuberta, and Sergio Escalera. Decorrelating neurons using persistence. In *NeurIPS 2023 Workshop on Symmetry and Geometry in Neural Representations*, volume 228 of *Proceedings of Machine Learning Research*, pages 164–182. PMLR, 2024a. URL https://proceedings.mlr.press/v228/ballester24a.html.

Rubén Ballester, Xavier Arnal Clemente, Carles Casacuberta, Meysam Madadi, Ciprian A. Corneanu, and Sergio Escalera. Predicting the generalization gap in neural networks using topological data analysis. *Neurocomputing*, 596:127787, 2024b. ISSN 0925-2312. URL https://www.sciencedirect.com/science/article/pii/S0925231224005587.

Serguei Barannikov, Ilya Trofimov, Nikita Balabin, and Evgeny Burnaev. Representation topology divergence: A method for comparing neural network representations. In Kamalika Chaudhuri, Stefanie Jegelka, Le Song, Csaba Szepesvari, Gang Niu, and Sivan Sabato, editors, *Proceedings of the 39th International Conference on Machine Learning*, volume 162 of *Proceedings of Machine Learning Research*, pages 1607–1626. PMLR, 17–23 Jul 2022. URL https://proceedings.mlr.press/v162/barannikov22a.html.

Abir Barbara, Younès Bennani, and Joseph Karkazan. On the use of persistent homology to control the generalization capacity of a neural network. In Biao Luo, Long Cheng, Zheng-Guang Wu, Hongyi Li, and Chaojie Li, editors, *Neural Information Processing*, pages 274–286, Singapore, 2024. Springer Nature Singapore. ISBN 978-981-99-8132-8.

Davis Blalock, Jose J. Gonzalez Ortiz, Jonathan Frankle, and John Guttag. What is the state of neural network pruning? In I. Dhillon, D. Papailiopoulos, and V. Sze, editors, *Proceedings of the 3rd MLSys Conference, Austin, TX, USA*, volume 2 of *Proceedings of Machine Learning and Systems*, pages 129–146, 2020. URL https://proceedings.mlsys.org/paper_files/paper/2020/file/6c44dc73014d66ba49b28d483a8f8b0d-Paper.pdf.

Thomas Bonis, Maks Ovsjanikov, Steve Oudot, and Frédéric Chazal. Persistence-based pooling for shape pose recognition. In Alexandra Bac and Jean-Luc Mari, editors, *Computational Topology in Image Context*, pages 19–29, Cham, 2016. Springer International Publishing. ISBN 978-3-319-39441-1.

Rickard Brüel Gabrielsson and Gunnar Carlsson. Exposition and interpretation of the topology of neural networks. In *2019 18th IEEE International Conference On Machine Learning And Applications (ICMLA)*, pages 1069–1076, 2019. https://doi.org/10.1109/ICMLA.2019.00180.

Gunnar Carlsson and Rickard Brüel Gabrielsson. Topological approaches to deep learning. In Nils A. Baas, Gunnar E. Carlsson, Gereon Quick, Markus Szymik, and Marius Thaule, editors, *Topological Data Analysis*, pages 119–146, Cham, 2020. Springer International Publishing. ISBN 978-3-030-43408-3.

Gunnar Carlsson, Tigran Ishkhanov, Vin de Silva, and Afra Zomorodian. On the local behavior of spaces of natural images. *International Journal of Computer Vision*, 76(1):1–12, Jan 2008. ISSN 1573-1405. URL https://doi.org/10.1007/s11263-007-0056-x.

Mathieu Carrière, Marco Cuturi, and Steve Oudot. Sliced Wasserstein kernel for persistence diagrams. In *Proceedings of the 34th International Conference on Machine Learning - Volume 70*, ICML'17, page 664–673. JMLR.org, 2017.

Ciprian A. Corneanu, Meysam Madadi, Sergio Escalera, and Aleix M. Martinez. What does it mean to learn in deep networks? And, how does one detect adversarial attacks? In *2019 IEEE/CVF*

Conference on Computer Vision and Pattern Recognition (CVPR), pages 4752–4761, 2019. https://doi.org/10.1109/CVPR.2019.00489.

Ciprian A. Corneanu, Sergio Escalera, and Aleix M. Martinez. Computing the testing error without a testing set. In *2020 IEEE/CVF Conference on Computer Vision and Pattern Recognition (CVPR)*, pages 2674–2682, 2020. https://doi.org/10.1109/CVPR42600.2020.00275.

Asim Kumar Debnath, Rosa L. Lopez de Compadre, Gargi Debnath, Alan J. Shusterman, and Corwin Hansch. Structure-activity relationship of mutagenic aromatic and heteroaromatic nitro compounds. correlation with molecular orbital energies and hydrophobicity. *Journal of Medicinal Chemistry*, 34(2):786–797, 1991. URL https://doi.org/10.1021/jm00106a046.

Jacob Devlin, Ming-Wei Chang, Kenton Lee, and Kristina Toutanova. BERT: Pre-training of deep bidirectional transformers for language understanding. In Jill Burstein, Christy Doran, and Thamar Solorio, editors, *Proceedings of the 2019 Conference of the North American Chapter of the Association for Computational Linguistics: Human Language Technologies, NAACL-HLT 2019, Minneapolis, MN, USA, June 2–7, 2019, Volume 1 (Long and Short Papers)*, pages 4171–4186. Association for Computational Linguistics, 2019. URL https://doi.org/10.18653/v1/n19-1423.

Barbara Di Fabio and Massimo Ferri. Comparing persistence diagrams through complex vectors. In Vittorio Murino and Enrico Puppo, editors, *Image Analysis and Processing – ICIAP 2015*, pages 294–305, Cham, 2015. Springer International Publishing. ISBN 978-3-319-23231-7.

Maxime Gabella. Topology of learning in feedforward neural networks. *IEEE Transactions on Neural Networks and Learning Systems*, 32(8):3588–3592, 2021. https://doi.org/10.1109/TNNLS.2020.3015790.

Yuri Gardinazzi, Giada Panerai, Karthik Viswanathan, Alessio Ansuini, Alberto Cazzaniga, and Matteo Biagetti. Persistent topological features in large language models, 2024. URL https://arxiv.org/abs/2410.11042.

Thomas Gebhart, Paul Schrater, and Alan Hylton. Characterizing the shape of activation space in deep neural networks. In *2019 18th IEEE International Conference on Machine Learning and Applications (ICMLA)*, pages 1537–1542, 2019. https://doi.org/10.1109/ICMLA.2019.00254.

Leander Girrbach, Anders Christensen, Ole Winther, Zeynep Akata, and Almut Sophia Koepke. Caveats of neural persistence in deep neural networks. In *2nd Annual TAG in Machine Learning*, 2023.

Morgane Goibert, Elvis Dohmatob, and Thomas Ricatte. An adversarial robustness perspective on the topology of neural networks. In *NeurIPS ML Safety Workshop*, 2022. URL https://openreview.net/forum?id=EtGd7pF237i.

Kaiming He, Xiangyu Zhang, Shaoqing Ren, and Jian Sun. Deep residual learning for image recognition. In *2016 IEEE Conference on Computer Vision and Pattern Recognition (CVPR)*, pages 770–778, 2016. https://doi.org/10.1109/CVPR.2016.90.

Jay A. Hennig, Matthew D. Golub, Peter J. Lund, Patrick T. Sadtler, Emily R. Oby, Kristin M. Quick, Stephen I. Ryu, Elizabeth C. Tyler-Kabara, Aaron P. Batista, Byron M. Yu, and Steven M. Chase. Constraints on neural redundancy. *eLife*, 7:e36774, Aug 2018. ISSN 2050-084X. URL https://doi.org/10.7554/eLife.36774.

Yiding Jiang, Pierre Foret, Scott Yak, Daniel M. Roy, Hossein Mobahi, Gintare Karolina Dziugaite, Samy Bengio, Suriya Gunasekar, Isabelle Guyon, and Behnam Neyshabur. Neurips 2020 Competition: Predicting generalization in deep learning, 2020. URL https://arxiv.org/abs/2012.07976.

Markelle Kelly, Rachel Longjohn, and Kolby Nottingham. The UCI Machine Learning Repository. https://archive.ics.uci.edu, 1987.

Nikita Klyuchnikov, Ilya Trofimov, Ekaterina Artemova, Mikhail Salnikov, Maxim Fedorov, and Evgeny Burnaev. NAS-Bench-NLP: Neural architecture search benchmark for natural language processing, 2020. URL https://arxiv.org/abs/2006.07116.

Alex Krizhevsky, Ilya Sutskever, and Geoffrey E Hinton. Imagenet classification with deep convolutional neural networks. In F. Pereira, C.J. Burges, L. Bottou, and K.Q. Weinberger, editors, *Advances in Neural Information Processing Systems*, volume 25. Curran Associates, Inc., 2012. URL https://proceedings.neurips.cc/paper_files/paper/2012/file/c399862d3b9d6b76c8436e924a68c45b-Paper.pdf.

Yann Lecun, Léon Bottou, Yoshua Bengio, and Patrick Haffner. Gradient-based learning applied to document recognition. *Proceedings of the IEEE*, 86(11):2278–2324, 1998. https://doi.org/10.1109/5.726791.

Ephy Love, Benjamin Filippenko, Vasileios Maroulas, and Gunnar E. Carlsson. Topological convolutional neural networks. In *TDA & Beyond*, 2020. URL https://openreview.net/forum?id=hntbh8Zo1V.

Daniel Lütgehetmann, Dejan Govc, Jason P. Smith, and Ran Levi. Computing persistent homology of directed flag complexes. *Algorithms*, 13(1), 2020. ISSN 1999-4893. URL https://www.mdpi.com/1999-4893/13/1/19.

Matthew Muller, Steve Kroon, and Stephan Chalup. Topological dynamics of functional neural network graphs during reinforcement learning. In Biao Luo, Long Cheng, Zheng-Guang Wu, Hongyi Li, and Chaojie Li, editors, *Neural Information Processing*, pages 190–204, Singapore, 2024. Springer Nature Singapore. ISBN 978-981-99-8138-0.

Gregory Naitzat, Andrey Zhitnikov, and Lek-Heng Lim. Topology of deep neural networks. *J. Mach. Learn. Res.*, 21(1), Jan 2020. ISSN 1532-4435.

National Institute of Standards and Technology. NIST TrojAI Competition Dataset. https://pages.nist.gov/trojai/docs/index.html#round-1, 2019.

Chris Olah, Alexander Mordvintsev, and Ludwig Schubert. Feature visualization. *Distill*, 2017. https://doi.org/10.23915/distill.00007. https://distill.pub/2017/feature-visualization.

Chris Olah, Arvind Satyanarayan, Ian Johnson, Shan Carter, Ludwig Schubert, Katherine Ye, and Alexander Mordvintsev. The building blocks of interpretability. *Distill*, 2018. https://distill.pub/2018/building-blocks.

Emilie Purvine, Davis Brown, Brett Jefferson, Cliff Joslyn, Brenda Praggastis, Archit Rathore, Madelyn Shapiro, Bci Wang, and Youjia Zhou. Experimental observations of the topology of convolutional neural network activations. In *Proceedings of the Thirty-Seventh AAAI Conference on Artificial Intelligence and Thirty-Fifth Conference on Innovative Applications of Artificial Intelligence and Thirteenth Symposium on Educational Advances in Artificial Intelligence*, AAAI'23/IAAI'23/EAAI'23. AAAI Press, 2023. ISBN 978-1-57735-880-0. URL https://doi.org/10.1609/aaai.v37i8.26134.

Archit Rathore, Nithin Chalapathi, Sourabh Palande, and Bei Wang. TopoAct: Visually exploring the shape of activations in deep learning. *Computer Graphics Forum*, 40(1):382–397, 2021. URL https://onlinelibrary.wiley.com/doi/abs/10.1111/cgf.14195.

Archit Rathore, Yichu Zhou, Vivek Srikumar, and Bei Wang. TopoBERT: Exploring the topology of fine-tuned word representations. *Information Visualization*, 22(3):186–208, 2023. URL https://doi.org/10.1177/14738716231168671.

Bastian Rieck, Matteo Togninalli, Christian Bock, Michael Moor, Max Horn, Thomas Gumbsch, and Karsten Borgwardt. Neural persistence: A complexity measure for deep neural networks using algebraic topology. In *International Conference on Learning Representations*, 2019. URL https://openreview.net/forum?id=ByxkijC5FQ.

Christian Szegedy, Wei Liu, Yangqing Jia, Pierre Sermanet, Scott Reed, Dragomir Anguelov, Dumitru Erhan, and Andrew Rabinovich. Going deeper with convolutions. In *2015 IEEE Conference on Computer Vision and Pattern Recognition (CVPR)*, pages 1–9, Los Alamitos, CA, USA, Jun 2015. IEEE Computer Society. URL https://doi.ieeecomputersociety.org/10.1109/CVPR.2015.7298594.

Satoru Watanabe and Hayato Yamana. Deep neural network pruning using persistent homology. In *2020 IEEE Third International Conference on Artificial Intelligence and Knowledge Engineering (AIKE)*, pages 153–156, 2020. https://doi.org/10.1109/AIKE48582.2020.00030.

Satoru Watanabe and Hayato Yamana. Overfitting measurement of convolutional neural networks using trained network weights. *International Journal of Data Science and Analytics*, 14(3):261–278, Sep 2022a. ISSN 2364-4168. URL https://doi.org/10.1007/s41060-022-00332-1.

Satoru Watanabe and Hayato Yamana. Topological measurement of deep neural networks using persistent homology. *Annals of Mathematics and Artificial Intelligence*, 90(1):75–92, Jan 2022b. ISSN 1573-7470. URL https://doi.org/10.1007/s10472-021-09761-3.

Matthew Wheeler, Jose Bouza, and Peter Bubenik. Activation landscapes as a topological summary of neural network performance. In *2021 IEEE International Conference on Big Data (Big Data)*, pages 3865–3870, Los Alamitos, CA, USA, dec 2021. IEEE Computer Society. https://doi.org/10.1109/BigData52589.2021.9671368. URL https://doi.ieeecomputersociety.org/10.1109/BigData52589.2021.9671368.

Méziane Yacoub and Younès Bennani. HVS: A heuristic for variable selection in multilayer artificial neural network classifier. *HAL*, 1997.

Lei Yang, Mengxue Xu, and Yunan He. Unraveling convolution neural networks: A topological exploration of kernel evolution. *Applied Sciences*, 14(5), 2024. ISSN 2076-3417. URL https://www.mdpi.com/2076-3417/14/5/2197.

Ben Zhang and Hongwei Lin. Functional loops: Monitoring functional organization of deep neural networks using algebraic topology. *Neural Networks*, 174:106239, 2024. ISSN 0893-6080. https://doi.org/10.1016/j.neunet.2024.106239. URL https://www.sciencedirect.com/science/article/pii/S0893608024001631.

Yang Zhao and Hao Zhang. Quantitative performance assessment of CNN units via topological entropy calculation. In *International Conference on Learning Representations*, 2022. URL https://openreview.net/forum?id=xFOyMwWPkz.

Songzhu Zheng, Yikai Zhang, Hubert Wagner, Mayank Goswami, and Chao Chen. Topological detection of trojaned neural networks. In A. Beygelzimer, Y. Dauphin, P. Liang, and J. Wortman Vaughan, editors, *Advances in Neural Information Processing Systems*, 2021. URL https://openreview.net/forum?id=1r2EannVuIA.

Youjia Zhou, Yi Zhou, Jie Ding, and Bei Wang. Visualizing and analyzing the topology of neuron activations in deep adversarial training. In *Topological, Algebraic and Geometric Learning Workshops 2023*, pages 134–145. PMLR, 2023.

Chapter 7
Training Dynamics and Loss Functions

Abstract In this chapter, we discuss four articles that focus on studying properties of the training process. The first one deals with the loss function used to train a network. The other three are focused on the evolution of weights during the training process and how the fractal dimension of weight trajectories is related to the generalization capacity of a network.

7.1 Loss Functions

One of the fundamental problems in deep learning is, given a learning problem, an architecture a, a loss function $\mathcal{L}$, and a training algorithm $\mathcal{A}$, to determine if the training algorithm $\mathcal{A}$ is capable of finding a neural network $\mathcal{N}$ with architecture a that minimizes the empirical risk $\widehat{\mathcal{R}}_{\mathcal{D}_{\text{train}}}$ for the learning problem. Usual training algorithms perform a gradient descent, following a path in the space of parameters of the neural network. This path is then heavily affected by the connectivity of the loss graph and by the presence of *local valleys*, in which the gradient descent algorithm can get stuck.

Nguyen (2019) studied a generalization of this problem for general FCFNNs depending on their activation functions, their graph structure (depth and width of layers), and the *shape* of the training dataset. This study focused on the number of connected components and the existence of local valleys on the graph of optimization targets

$$\mathcal{L}(\theta) = f\big(\phi_{\mathcal{N}_\theta}(x_1), \ldots, \phi_{\mathcal{N}_\theta}(x_m)\big),$$

where f is a convex function, and $\mathcal{D}_{\text{train}} = \{(x_i, y_i)\}_{i=1}^{m}$. Such generalized optimization targets allow training algorithms to minimize any convex function with respect to the parameters of neural networks that could be useful to obtain better parameters for the network, not restricting the minimization to empirical risks.

Although this is a more complex scenario than the one presented in Sect. 2.1, empirical risk functions for many loss functions, such as categorical cross entropy, are examples of these types of optimization targets. For the discussion of this article,

© The Author(s), under exclusive license to Springer Nature Switzerland AG 2026

R. Ballester et al., *Topological Data Analysis for Neural Networks*, SpringerBriefs in Computer Science, https://doi.org/10.1007/978-3-032-08283-1_7

we assume that the activation functions for all layers except for the last one are the same, and that the last-layer activation function is the identity.

The first insight on the optimization target graph is that, for strictly monotonic activation functions φ with $\text{Im}(\varphi) = \mathbb{R}$, those FCFNNs with widths strictly decreasing layer by layer, that is, $N_i > N_{i+1}$ for $i \in [L-1]$, with at least two non-input layers, i.e., $L \geq 2$, and a linearly independent training dataset $\mathcal{D}_{\text{train}}$, every sublevel set of $\mathcal{L}$ —that is, the sets $\mathcal{L}^{-1}(-\infty, \alpha]$ for $\alpha \in \mathbb{R}$— is connected and also every non-empty connected component of every level set $\mathcal{L}^{-1}(\alpha)$ is unbounded. Connectedness of sublevel sets implies a well-behaved optimization target function, and unboundedness of level sets implies that there are no local valleys in the optimization target graph, understanding by a local valley a non-empty connected component of some strict sublevel set $\mathcal{L}^{-1}(-\infty, \alpha)$.

A bad local valley is a local valley in which the target $\mathcal{L}$ cannot be arbitrarily close to the lower bound of the convex function f inducing the target function, which is also a lower bound of $\mathcal{L}$ but does not depend on any concrete neural network. Bad local valleys are harmful to the optimization problem because if the training process enters them when optimizing the target function, the parameters obtained at the end of the learning process are provably not optimal.

Nguyen (2019) proved that, for activation functions as before that also satisfy that there are no nonzero coefficients $(\lambda_i, a_i)_{i=1}^{P}$ with $a_i \neq a_j$ for all $i \neq j$ such that

$$\varphi(x) = \sum_{i=1}^{p} \lambda_i \, \varphi(x - a_i)$$

for all $x \in \mathbb{R}$, those FCFNNs with a layer $l \in [L-1]$ satisfying $N_l > |\mathcal{D}_{\text{train}}|$ and $N_i > N_{i+1}$ for $i \in \{l+1, \ldots, L\}$ induce target functions $\mathcal{L}$ with no bad local valleys and, if $l \leq L-2$, then every local valley of $\mathcal{L}$ is unbounded. This implies that, from any initial parameter, there is a continuous path on which the loss function is nonincreasing from it to a point that is arbitrarily close to the infimum of the loss.

As we have seen so far, widths of neural network layers play a significant role in the configuration of the optimization target graph. The first hidden layer always plays an important role in this configuration, since it determines the first transformation of the data into some space from which features from the inputs are extracted. Assuming that the activation functions of the network satisfy the two previous assumptions, and assuming that $N_1 > 2|\mathcal{D}_{\text{train}}|$ and that $N_i > N_{i+1}$ for $i \in \{2, \ldots, L-1\}$, Nguyen (2019) proved that every sublevel set of $\mathcal{L}$ is connected and also that every connected component of every level set of $\mathcal{L}$ is unbounded. This is a stronger result than the previous one, as it implies that not only there are no bad local valleys but also there is a unique global valley.

Many current activation functions, such as the leaky ReLU, satisfy the previous assumptions. However, the usual ReLU does not, and further assumptions are needed to have a good behavior of the target graph. Letting $N_{\text{min}} = \min_{i \in [L-1]} N_i$, for FCFNNs with activation functions φ only satisfying the above assumption involving coefficients $(\lambda_i, a_i)_{i=1}^{p}$, if $N_{\min} > |\mathcal{D}_{\text{train}}|$, then $\mathcal{L}$ has no bad local valleys and, if $N_{\min} > 2|\mathcal{D}_{\text{train}}|$, then every sublevel set of $\mathcal{L}$ is connected.

Further exploring the interplay between parameter space geometry and training dynamics, Nurisso et al. (2024) performed a similar study focusing on neural networks with a single hidden layer where $\varphi^{(1)}$ is the identity and $\varphi^{(2)}$ satisfies some technical conditions met by standard activations such as ReLU and leaky ReLU. The study showed that gradient flow trajectories —which are weight trajectories of a continuous approximation of the gradient descent— are confined to a subset of the parameter space given by the intersection of quadric hypersurfaces dependent on the initialization parameters of the networks when minimizing the empirical risk for a loss function $\mathcal{L}(f, x, y) = \ell(f(x), y)$ with ℓ differentiable. Nurisso et al. (2024) fully characterized the Betti numbers of this set, with a particular emphasis on its connectivity (the zeroth Betti number). Their analysis demonstrated that this set can be disconnected depending on the initialization, thereby generating topological obstructions that may prevent gradient flow from reaching a global optimum of the loss. Furthermore, it was shown how to generate initializations for the network that result in this invariant set having only one connected component and how weights yielding equivalent network functions ϕ_N interplay with the different connected components, accompanied by an experimental study of their findings on toy datasets.

Some of the previous assumptions are too strong to be satisfied in many practical scenarios. However, these results serve as a basis for a better understanding of the shape of the target function. A similar approach using tools like persistent homology or Mapper to capture the connectivity and shape of optimization target graphs for modern neural networks could be a promising research line. This could help verify if the claims made in these cases can be extrapolated to more general scenarios.

7.2 Fractal Dimension of Weight Trajectories

The last approach that we discuss in this survey is related to the evolution of network parameters during the training process. In this case, training algorithms $\mathcal{A}$ are assumed to be continuous, which means that parameters evolve over a time period $[0, T] \subseteq \mathbb{R}$ for which each instant of time $t \in [0, T]$ has an associated weight value θ_t. Although this is not the case for real scenarios, since computers work in the discrete domain, there are good continuous approximations of classical *discrete* optimization algorithms such as gradient descent that allow us to study these discrete processes with properties of the continuous approximations.

Simsekli et al. (2020) discovered that the generalization capacity of neural networks was connected with the heavy-tailed behavior of weight trajectories $\Theta_{\mathcal{A}} = \{\theta_t : t \in [0, T]\}$ generated during training, in particular, with its upper-box dimension, a dimensionality measure for fractals. A difficulty with this link is that many strong assumptions about the training algorithm and the space of weights were needed to formally prove the connection.

Birdal et al. (2021) relaxed these assumptions by cleverly using a previous result that connected fractal dimension with persistent homology (Kozma et al. 2006; Schweinhart 2021). This result says that, for $\Theta \subseteq \mathbb{R}^d$ a bounded set, we have that

$$\dim_{\text{PH}}\Theta = \dim_{\text{PH}}^0\Theta = \dim_{\text{Box}}\Theta,$$

where $\dim_{\text{Box}}$ is the upper-box fractal dimension, and $\dim_{\text{PH}}^k$ is the persistent homology dimension given by

$$\dim_{\text{PH}}^k\Theta = \inf\left\{\alpha : E_\alpha^k(\Theta_{<\infty}) < C : \exists C > 0 \text{ for all finite } \Theta_{<\infty} \subseteq \Theta\right\},$$

$$E_\alpha^k(P) = \sum_{(b,d)\in D} |d - b|^\alpha, \qquad D = D(\mathbb{V}_k(\text{VR}(P, \|\cdot\|_2))),$$

that is, the infimum of the exponents α for which E_α^k is uniformly bounded for all finite subsets $\Theta_{<\infty}$.

In particular, Birdal et al. (2021) proved that, given any compact set of possible weights Θ, for example, the weight training trajectories $\Theta_{\mathcal{A}}$, and a loss function $\mathcal{L}$ bounded by B and Lipschitz continuous on the set of possible parameters θ of a fixed neural network architecture, then, for a sufficiently large size m of the training dataset, the following holds:

$$\sup_{\theta\in\Theta} \left|\widehat{\mathcal{R}}_{\mathcal{D}_{\text{train}}}(\phi_{N_\theta}) - \mathcal{R}(\phi_{N_\theta})\right| \leq 2B \sqrt{\frac{(\dim_{\text{PH}}\Theta + 1)\log^2(mL^2)}{m} + \frac{\log(7M/\gamma)}{m}}$$

with probability at least $1 - \gamma$ over a training dataset $\mathcal{D}_{\text{train}}$ with m elements sampled i.i.d. from the data distribution, where N_θ denotes the neural network N defined with the fixed architecture and parameters θ, and M denotes a constant depending on some technical assumptions about the objects involved in the bound.

This bound uses persistent homology dimension, which cannot be computed exactly by a computer program. Birdal et al. (2021) propose an algorithm to estimate this quantity for finite sets of weights. Estimations of persistent homology dimension were found to be significantly correlated with the generalization gap calculated as the difference between the accuracy in train and the accuracy in test, indicating that lower persistent homology dimensions were associated with better generalization capacities of networks for a wide variety of networks, including simple FCFNN models and AlexNet, trained on MNIST, CIFAR-10, and CIFAR-100 datasets.

Thanks to the differentiability theory for persistence diagrams, Birdal et al. (2021)were able to develop a regularization term with the objective of minimizing the above bound and, thus, improving the generalization of regularized trained networks. This is exactly what happened for a LeNet-5 (Lecun et al. 1998) architecture trained on CIFAR-10 in the experiments performed by Birdal et al. (2021), especially for training procedures that failed to converge to a good set of parameters only by themselves.

Bounds were improved in a follow-up paper by Dupuis et al. (2023), in which the persistent homology dimension was used again to compute fractal dimensions. As the methods employed to improve bounds are not related with TDA, we do not analyze the paper in this monograph, although the results are insightful and interesting in their own right.

More recently, Tan et al. (2024) questioned the correlation between persistent homology dimension measures, as calculated by Birdal et al. (2021), and the generalization capabilities of neural networks through a comprehensive set of experiments. Although limited on the diversity of architectures and datasets, the results suggest that the persistent homology dimension of weight trajectories is not sufficient to explain generalization in neural networks, and that most probably other factors are involved in the generalization capabilities of neural networks, which might be captured by other methods.

In parallel, Andreeva et al. (2024) expanded upon the relationship between generalization and persistent homology of training weight trajectories, focusing on discrete training algorithms. Their study combined theoretical and empirical approaches. Theoretically, they developed new upper bounds for the generalization gap. Empirically, they conducted an extensive evaluation assessing the relationship between various topological and geometric measures and the generalization gap. To the best of our knowledge, that work represents the first application of topological data analysis (TDA) to examine generalization in contemporary, practically relevant models such as transformers and graph neural networks.

References

Rayna Andreeva, Benjamin Dupuis, Rik Sarkar, Tolga Birdal, and Umut Şimşekli. Topological generalization bounds for discrete-time stochastic optimization algorithms, 2024. URL https://arxiv.org/abs/2407.08723.

Tolga Birdal, Aaron Lou, Leonidas J. Guibas, and Umut Simsekli. Intrinsic dimension, persistent homology and generalization in neural networks. In M. Ranzato, A. Beygelzimer, Y. Dauphin, P. S. Liang, and J. Wortman Vaughan, editors, *Advances in Neural Information Processing Systems*, volume 34, pages 6776–6789. Curran Associates, Inc., 2021. URL https://proceedings.neurips.cc/paper_files/paper/2021/file/35a12c43227f217207d4e06ffefe39d3-Paper.pdf.

Benjamin Dupuis, George Deligiannidis, and Umut Simsekli. Generalization bounds using data-dependent fractal dimensions. In Andreas Krause, Emma Brunskill, Kyunghyun Cho, Barbara Engelhardt, Sivan Sabato, and Jonathan Scarlett, editors, *Proceedings of the 40th International Conference on Machine Learning*, volume 202 of *Proceedings of Machine Learning Research*, pages 8922–8968. PMLR, 23–29 Jul 2023. URL https://proceedings.mlr.press/v202/dupuis23a.html.

Gady Kozma, Zvi Lotker, and Gideon Stupp. The minimal spanning tree and the upper box dimension. *Proceedings of the American Mathematical Society*, 134(4):1183–1187, 2006. ISSN 00029939, 10886826.

Yann Lecun, Léon Bottou, Yoshua Bengio, and Patrick Haffner. Gradient-based learning applied to document recognition. *Proceedings of the IEEE*, 86(11):2278–2324, 1998. doi: 10.1109/5.726791.

Quynh Nguyen. On connected sublevel sets in deep learning. In Kamalika Chaudhuri and Ruslan Salakhutdinov, editors, *Proceedings of the 36th International Conference on Machine Learning*, volume 97 of *Proceedings of Machine Learning Research*, pages 4790–4799. PMLR, 09–15 Jun 2019. URL https://proceedings.mlr.press/v97/nguyen19a.html.

Marco Nurisso, Pierrick Leroy, and Francesco Vaccarino. Topological obstruction to the training of shallow ReLU neural networks. In *The Thirty-eighth Annual Conference on Neural Information Processing Systems*, 2024. URL https://openreview.net/forum?id=3hcn0UxP72.

Benjamin Schweinhart. Persistent homology and the upper box dimension. *Discrete & Computational Geometry*, 65(2):331–364, Mar 2021. ISSN 1432-0444. URL https://doi.org/10.1007/s00454-019-00145-3.

Umut Simsekli, Ozan Sener, George Deligiannidis, and Murat A. Erdogdu. Hausdorff dimension, heavy tails, and generalization in neural networks. In H. Larochelle, M. Ranzato, R. Hadsell, M.F. Balcan, and H. Lin, editors, *Advances in Neural Information Processing Systems*, volume 33, pages 5138–5151. Curran Associates, Inc., 2020. URL https://proceedings.neurips.cc/paper_files/paper/2020/file/37693cfc748049e45d87b8c7d8b9aacd-Paper.pdf.

Charlie Tan, Inés García-Redondo, Qiquan Wang, Michael M. Bronstein, and Anthea Monod. On the limitations of fractal dimension as a measure of generalization, 2024. URL https://arxiv.org/abs/2406.02234.

Chapter 8
Challenges, Future Directions, and Conclusions

Abstract This final chapter provides a summary and discussion of the main facts presented in the previous chapters and outlines potential directions for future research, such as persistent path homology of weighted graphs, topology-based regularization of generative models, and multiparameter persistence.

In this survey we have seen many examples of how topological data analysis, and particularly persistent homology and Mapper, can be applied to study the properties of neural networks. In Sect. 5.1, we saw how to compute two homologies that yielded different information about a neural network only by taking its graph without weights. However, we noticed that the ranks of those homology groups were invariant to some important structural properties of the neural network, such as the order of the layers. Also, their values were simply a combination of the width and depth of the graph, not capturing much information about the network.

To obtain a better insight into the properties of neural networks, more fine-grained homology theories for directed graphs are needed. A promising line of work would be to study the persistent path homology (Chowdhury and Mémoli 2018) of graphs augmented with weights associated to each edge, as in Chap. 6.

In Chap. 5, we saw how topology is useful in recovering the *topology* of decision regions and boundaries, with applications especially in model selection. Furthermore, we saw that GTDA, an evolution of the Mapper algorithm for graphs as input, has been successful in analyzing patterns in output spaces of neural networks, leading to a better understanding of misclassification errors in several datasets.

Thanks to the differentiability theory of persistence diagrams, we saw how to *improve* decision regions by *simplifying* them. In addition, we discussed how topology is useful for capturing the quality —measured in different ways— of generative models. Particularly interesting is the study of the disentanglement of generative models, since being able to decouple the different sources of variation could lead to a better control and quality of the networks' outputs. Refining the article by Zhou et al. (2021) and using its measures to regularize generative neural

© The Author(s), under exclusive license to Springer Nature Switzerland AG 2026

R. Ballester et al., *Topological Data Analysis for Neural Networks*, SpringerBriefs in Computer Science, https://doi.org/10.1007/978-3-032-08283-1_8

networks could be an interesting approach to improving neural network design using TDA.

In Chap. 6, we analyzed methods for studying neural network parameters and activations. We split the chapter into those articles using Mapper and those using persistent homology, because of the differences in their approaches. We saw how Carlsson and Brüel Gabrielsson (2020) found very interesting connections between the topologies of the space of natural images and the space of weights of some layers in convolutional networks, among many other interesting topology configurations of the convolutional weights. We also saw how Gabella (2021) studied Mapper graphs of the evolution of weights during training, finding a very interesting pattern in the weight distribution of one of the architectures tried in their experiments: the weights seemed to be distributed in a *surface*. A closer look at the topology of the weight evolution for a higher variety of architectures must be taken to realize if the weights share common patterns in similar problems, such as surface arrangement. For activations, we found that the general approach proposed by TopoAct (Rathore et al. 2021) was extremely useful for understanding the internal behavior of neural networks, with the many ramifications presented in the monograph.

Concerning persistent homology papers, we could broadly classify the approaches depending on either the set of neurons/edges analyzed (layer by layer versus all the neurons), or the dissimilarity strategy (weights versus distances between activations or versus a combination of weights and activations).

In many cases, there was a connection between the different topological summaries extracted from Vietoris–Rips persistence diagrams and the properties of neural networks, especially with their generalization capacities. However, Girrbach et al. (2023) showed that most of the information extracted by one of the most influential articles in this chapter could be captured by taking simpler, non-topological summaries of the neural network. Although this does not mean that TDA does not provide unique information about the internal workings of the neural networks, we think that more ablation studies must be performed in the articles claiming the utility of TDA for the analysis of neural networks, as almost all the work performed in this area is experimental, and, usually, there is no theory supporting the hypotheses connecting topological summaries and neural network properties.

Another interesting discussion in Chap. 6 is the opposition of views on the evolution of the topology of data through the different layers. In most works, such as the one by Naitzat et al. (2020), it is stated and verified that the topology of data is simplified by the action of layers. However, work by Wheeler et al. (2021) arrived to the opposite conclusion. This contradiction supports the necessity of performing more diverse studies.

This circumstance leads to one of the fundamental limitations of Chap. 6: most of the experiments are performed on classical CNN and FCFNN architectures and do not say anything about more modern architectures, like transformers. To be useful, TDA must be applied to state-of-the-art architectures, not only as a proof of concept on *simple* neural networks. This is a fundamental step for gaining the trust of the non-TDA deep learning community in TDA methods. Also, since many people working in this area are mathematicians, it would be desirable that TDA methods be accompanied by theoretical results, as in Birdal et al. (2021).

In Chap. 7, we saw how TDA was used to study loss functions and training weights. In this case, we observed that persistence diagrams, which come originally from topology, were used to compute fractal dimensions, that belong to the geometry realm. This is an exciting result because it means that persistent homology can be used not only to infer the topology of data, but also to obtain geometric properties. We think that seeing persistent homology as a tool that also extracts local information from data could be useful in performing new work to analyze network structures.

Probably the most critical drawback of topological data analysis for neural networks is the high computational complexity in time and memory of computing invariants of persistence modules, like persistence diagrams. For dimension zero, algorithms based on the minimum spanning tree (Chazelle 2000) or the single linkage algorithm (Sibson 1973), have time complexities $O(e \cdot \alpha(e, n))$ and $O(n^2)$, respectively, where α is the very slow growing functional inverse of the Ackermann function Tarjan (1975), and e is the number of edges of the simplicial complex, whose cardinality is $O(n^2)$. Also, for single-linkage clustering, the memory complexity is linear, that is, $O(n)$. This makes these algorithms suitable for some of the applications for which the number of points is relatively small, such as the analysis of weights or activations layer by layer. However, for bigger point clouds, such as those in problems in which we use the whole set of neurons or weights, these algorithms may become infeasible for modern big neural networks. The situation is worse for dimensions greater than or equal to one, where typical persistence diagram computation algorithms have a time and memory complexity of $O(n^w)$ and $O(n^2)$, respectively, where w is the matrix multiplication exponent (currently $w < 2.4$) and n is the number of simplices generated through the filtration of simplicial complexes (Birdal et al. 2021).

Computational problems motivate the development of more suitable algorithms to compute persistence diagrams or alternative invariants that also capture information of neural networks. Although powerful tools for computing Vietoris–Rips persistence diagrams are available, such as Ripser (Bauer 2021), Gudhi (The GUDHI Project 2015), SLINK (Sibson 1973), algorithms that parallelize computations, such as Ripser++ (Zhang et al. 2020) or giotto-ph (Pérez et al. 2021), are of special interest due to the increasing availability of powerful hardware and software methods to perform distributed computing. Another possible approach to tackle complexity would be developing approximations of known filtrations the persistent homology of which is efficient to compute, as in Sheehy (2013), or directly learning persistent homology features with machine learning models, as in Montufar et al. (2020).

As we have seen in this survey, there are many different ways to produce meaningful filtrations for neural networks in all the domains we reviewed. For this reason, it would be desirable to capture the joint information of some of them at the same time. For some specific problems, like analyzing the evolution of data distribution through the layers, filtrations by distances alone may not be good enough, as they do not consider important data properties such as the density of points in a sample.

Luckily, *multiparameter persistence* (Botnan and Lesnick 2023) deals with this kind of problem setting, allowing to extract information from non-linear filtrations, that is, with more than one varying parameter. However, unluckily, multiparameter persistence does not have a canonical, easy-to-use and computationally efficient representation like persistence diagrams for one-dimensional persistence. There are already works proposing a rich body of useful, and even differentiable, representations for multiparameter persistence —see, for example Loiseaux et al. (2023a), Loiseaux et al. (2023b), and Xin et al. (2023); Scoccola et al. (2024).

We view multiparameter persistence as one of the main lines of work not only for the analysis of deep neural networks, but also for the whole machine learning community, since multiparameter persistence provides sharper ways to extract information from data. However, we are far from having usable representations for real neural network use cases due to the huge quantity of points in real datasets and neurons in architectures. For this reason, additional fundamental research on the topic is needed before we can grasp its advantages in the deep learning community. Further work in this direction could be based on three basic pillars:

1. Ease of use of multiparameter representations, since representations must be intuitive for the average machine learning researcher;
2. Efficiency of computation;
3. Differentiability of representations with respect to the point clouds, to allow regularization in learning tasks.

In conclusion, we have seen that the use of TDA applied to studying neural networks is an exciting path that can be useful in many scenarios. TDA, although computationally expensive, is a tool that has been shown to be connected with many interesting properties of neural networks such as generalization in classification problems or entanglement and latent space quality in generative models. Moreover, we have seen that there are many ways in which TDA can be used in applied scenarios, converting TDA into an essential tool for deep learning practitioners.

References

Ulrich Bauer. Ripser: Efficient computation of Vietoris–Rips persistence barcodes. *Journal of Applied and Computational Topology*, 5(3):391–423, Sep 2021. ISSN 2367-1734. URL https://doi.org/10.1007/s41468-021-00071-5.

Tolga Birdal, Aaron Lou, Leonidas J. Guibas, and Umut Simsekli. Intrinsic dimension, persistent homology and generalization in neural networks. In M. Ranzato, A. Beygelzimer, Y. Dauphin, P. S. Liang, and J. Wortman Vaughan, editors, *Advances in Neural Information Processing Systems*, volume 34, pages 6776–6789. Curran Associates, Inc., 2021. URL https://proceedings.neurips.cc/paper_files/paper/2021/file/35a12c43227f217207d4e06ffefe39d3-Paper.pdf.

Magnus Bakke Botnan and Michael Lesnick. An introduction to multiparameter persistence, 2023. URL https://arxiv.org/abs/2203.14289.

Gunnar Carlsson and Rickard Brüel Gabrielsson. Topological approaches to deep learning. In Nils A. Baas, Gunnar E. Carlsson, Gereon Quick, Markus Szymik, and Marius Thaule, editors, *Topological Data Analysis*, pages 119–146, Cham, 2020. Springer International Publishing. ISBN 978-3-030-43408-3.

Bernard Chazelle. A minimum spanning tree algorithm with inverse-Ackermann type complexity. *J. ACM*, 47(6):1028–1047, nov 2000. ISSN 0004-5411. URL https://doi.org/10.1145/355541.355562.

Samir Chowdhury and Facundo Mémoli. Persistent path homology of directed networks. In *Proceedings of the Twenty-Ninth Annual ACM-SIAM Symposium on Discrete Algorithms*, SODA 2018, page 1152–1169, USA, 2018. Society for Industrial and Applied Mathematics. ISBN 9781611975031.

Maxime Gabella. Topology of learning in feedforward neural networks. *IEEE Transactions on Neural Networks and Learning Systems*, 32(8):3588–3592, 2021. https://doi.org/10.1109/TNNLS.2020.3015790.

Leander Girrbach, Anders Christensen, Ole Winther, Zeynep Akata, and Almut Sophia Koepke. Caveats of neural persistence in deep neural networks. In *2nd Annual TAG in Machine Learning*, 2023.

David Loiseaux, Mathieu Carrière, and Andrew Blumberg. A framework for fast and stable representations of multiparameter persistent homology decompositions. In *Thirty-seventh Conference on Neural Information Processing Systems*, 2023a. URL https://openreview.net/forum?id=nKCUDd9GYu.

David Loiseaux, Luis Scoccola, Mathieu Carrière, Magnus Bakke Botnan, and Steve Oudot. Stable vectorization of multiparameter persistent homology using signed barcodes as measures. In *Thirty-seventh Conference on Neural Information Processing Systems*, 2023b. URL https://openreview.net/forum?id=4mwORQjAim.

Guido Montufar, Nina Otter, and Yu Guang Wang. Can neural networks learn persistent homology features? In *TDA & Beyond*, 2020. URL https://openreview.net/forum?id=pqpXM1Wjsxe.

Gregory Naitzat, Andrey Zhitnikov, and Lek-Heng Lim. Topology of deep neural networks. *J. Mach. Learn. Res.*, 21(1), Jan 2020. ISSN 1532-4435.

Julián Burella Pérez, Sydney Hauke, Umberto Lupo, Matteo Caorsi, and Alberto Dassatti. giotto-ph: A Python library for high-performance computation of persistent homology of Vietoris–Rips filtrations, 2021.

Archit Rathore, Nithin Chalapathi, Sourabh Palande, and Bei Wang. TopoAct: Visually exploring the shape of activations in deep learning. *Computer Graphics Forum*, 40(1):382–397, 2021. URL https://onlinelibrary.wiley.com/doi/abs/10.1111/cgf.14195.

Luis Scoccola, Siddharth Setlur, David Loiseaux, Mathieu Carrière, and Steve Oudot. Differentiability and optimization of multiparameter persistent homology. In *Forty-first International Conference on Machine Learning*, 2024. URL https://openreview.net/forum?id=ixdfvnO0uy.

Donald R. Sheehy. Linear-size approximations to the Vietoris–Rips filtration. *Discrete & Computational Geometry*, 49(4):778–796, Jun 2013. ISSN 1432-0444. URL https://doi.org/10.1007/s00454-013-9513-1.

Robin Sibson. SLINK: An optimally efficient algorithm for the single-link cluster method. *The Computer Journal*, 16(1):30–34, 01 1973. ISSN 0010-4620. URL https://doi.org/10.1093/comjnl/16.1.30.

Robert Endre Tarjan. Efficiency of a good but not linear set union algorithm. *J. ACM*, 22(2):215–225, apr 1975. ISSN 0004-5411. URL https://doi.org/10.1145/321879.321884.

The GUDHI Project. *GUDHI User and Reference Manual*. GUDHI Editorial Board, 2015. URL http://gudhi.gforge.inria.fr/doc/latest/.

Matthew Wheeler, Jose Bouza, and Peter Bubenik. Activation landscapes as a topological summary of neural network performance. In *2021 IEEE International Conference on Big Data (Big Data)*, pages 3865–3870, Los Alamitos, CA, USA, dec 2021. IEEE Computer Society. URL https://doi.ieeecomputersociety.org/10.1109/BigData52589.2021.9671368.

Cheng Xin, Soham Mukherjee, Shreyas N. Samaga, and Tamal K. Dey. GRIL: A 2-parameter persistence based vectorization for machine learning. In *Proceedings of 2nd Annual Workshop on*

Topology, Algebra, and Geometry in Machine Learning (TAG-ML), volume 221 of *Proceedings of Machine Learning Research*, pages 313–333. PMLR, 28 Jul 2023. URL https://proceedings. mlr.press/v221/xin23a.html.

Simon Zhang, Mengbai Xiao, and Hao Wang. GPU-Accelerated computation of Vietoris-Rips persistence barcodes. In *36th International Symposium on Computational Geometry (SoCG 2020)*. Schloss Dagstuhl-Leibniz-Zentrum für Informatik, 2020.

Sharon Zhou, Eric Zelikman, Fred Lu, Andrew Y. Ng, Gunnar E. Carlsson, and Stefano Ermon. Evaluating the disentanglement of deep generative models through manifold topology. In *International Conference on Learning Representations*, 2021. URL https://openreview.net/ forum?id=djwS0m4Ft_A.

Appendix

Peer-Reviewed Articles Discussed in This Book

See Table A.1.

Table A.1 Table containing the published articles reviewed in this survey ordered by year of publication. Each row corresponds to an article. The first column contains the title; the second column contains a one-line summary; the third column contains the categories associated to each article; and the last column contains the applications introduced in Chap. 4. The four possible categories are 1. Architecture; 2. Input and output spaces; 3. Internal representations and activations; 4. Training dynamics and loss functions. The seven possible applications are 1. Regularization; 2. Pruning of neural networks; 3. Detection of adversarial, out-of-distribution and shifted examples; 4. Detection of trojaned networks; 5. Model selection; 6. Prediction of accuracy; 7. Quality assessment of generative models

Title	Summary	Cat.	Apps.
On the complexity of neural network classifiers: A comparison between shallow and deep architectures (Bianchini and Scarselli 2014)	Bounds on the Betti numbers of the positive decision region generated by binary classification neural networks with Pfaffian activations	(2)	–
Topological approaches to deep learning (Carlsson and Brüel Gabrielsson 2020)	Topological analysis of the weights of convolutional neural networks and generalization of convolutional neural networks	(3)	–
Topological data analysis of decision boundaries with application to model selection (Ramamurthy et al. 2019)	Study and approximation of the topology of network decision boundaries	(2)	(5)
A topological regularizer for classifiers via persistent homology (Chen et al. 2019)	Regularization of neural networks by modifying their decision regions using differentiable persistent homology	(2)	(1)

Continued on next page

© The Author(s), under exclusive license to Springer Nature Switzerland AG 2026
R. Ballester et al., *Topological Data Analysis for Neural Networks*, SpringerBriefs in Computer Science, https://doi.org/10.1007/978-3-032-08283-1

Table A.1 (continued)

Title	Summary	Cat.	Apps.
Geometry score: A method for comparing generative adversarial networks (Khrulkov and Oseledets 2018)	Measurement of GAN quality using persistent homology	(2)	(7)
What does it mean to learn in deep networks? And, how does one detect adversarial attacks? (Corneanu et al. 2019)	Study of generalization in terms of the persistent homology of neuron activations and their correlations	(3)	(1, 3)
On connected sublevel sets in deep learning (Nguyen 2019)	Study of the connectivity, boundedness, and local minima of sublevel sets of convex losses for overparameterised neural networks with piecewise linear activation functions	(4)	–
Neural persistence: A complexity measure for deep neural networks using algebraic topology (Rieck et al. 2019)	Topological complexity measure for neural networks based on their weights	(3)	(1)
Characterizing the shape of activation space in deep neural networks (Gebhart et al. 2019)	Topological characterization of the neuron activations given a sample	(3)	(3)
Exposition and interpretation of the topology of neural networks (Brüel Gabrielsson and Carlsson 2019)	Topological analysis of the weights of convolutional neural networks and connection to generalization capacity of models	(3)	–
Path homologies of deep feedforward networks (Chowdhury et al. 2019)	Analysis of path and directed flag homology groups of MLP directed graphs	(1)	–
Topology of deep neural networks (Naitzat et al. 2020)	Study of the topology of the data through layer transformations	(3)	–
Finding the homology of decision boundaries with active learning (Li et al. 2020)	Use of active learning to improve the methods of topological data analysis of decision boundaries with application to model selection	(2)	(5)
Computing the testing error without a testing set (Corneanu et al. 2020)	Regression of test accuracy using vectorizations of persistence diagrams	(3)	(6)
Topological detection of trojaned neural networks (Zheng et al. 2021)	Study of trojaned networks in terms of the persistent homology of neuron activations and their correlations	(3)	(4)
Experimental stability analysis of neural networks in classification problems with confidence sets for persistence diagrams (Akai et al. 2021)	Study of persistence diagrams of last hidden layer activations and its connection with generalization	(3)	–
PHom-GeM: Persistent homology for generative models (Charlier et al. 2019)	Comparison between persistence diagrams of real and generated manifolds using generative models	(2)	(7)

Continued on next page

Table A.1 (continued)

Title	Summary	Cat.	Apps.
TopoAct: Visually exploring the shape of activations in deep learning (Rathore et al. 2021)	Analysis of the Mapper graph of neuron activations for each layer	(3)	–
Deep neural network pruning using persistent homology (Watanabe and Yamana 2020)	Pruning of neural networks using persistent homology	(3)	(2)
Intrinsic dimension, persistent homology and generalization in neural networks (Birdal et al. 2021)	Generalization error bounds using persistent homology dimension on the training weights	(3, 4)	(1)
Activation landscapes as a topological summary of neural network performance (Wheeler et al. 2021)	Study of the topology of the data through layer transformations and its connection with training accuracy	(3)	–
Topology of learning in feedforward neural networks (Gabella 2021)	Analysis of the evolution of the weights of a neural network during training using Mapper	(3)	–
Topological uncertainty: Monitoring trained neural networks through persistence of activation graphs (Lacombe et al. 2021)	Uncertainty measurement of neural network predictions using the topology of neuron activations	(3)	(3, 5)
Topological measurement of deep neural networks using persistent homology (Watanabe and Yamana 2022b)	Computation of persistence diagrams using neuron path relevance and connection with network expressivity and problem difficulty	(3)	–
Evaluating the disentanglement of deep generative models with manifold topology (Zhou et al. 2021)	Measurement of disentanglement of generative neural networks	(2)	–
Quantitative performance assessment of CNN units via topological entropy calculation (Zhao and Zhang 2022)	Measurement of quality of convolutions in a neural network using persistent homology	(3)	–
Overfitting measurement of deep neural networks using no data (Watanabe and Yamana 2022a)	Study of overfitting by analysing the points near the diagonal of a persistence diagram generated from the weights of a neural network	(3)	–
An adversarial robustness perspective on the topology of neural networks (Goibert et al. 2022)	Analysis of adversarial examples by means of the topology of the subgraph of under-optimized edges of neural networks	(3)	(3)
Representation topology divergence: A method for comparing neural network representations (Barannikov et al. 2022)	Definition of similarity between data representations using persistent homology	(3)	(7)
On the topological expressive power of neural networks (Petri and Leitão 2020)	Measurement of expressivity of network architectures	(2)	–
Generalization bounds using data-dependent fractal dimensions (Dupuis et al. 2023)	Data-dependent generalization error bounds using persistent homology dimension on the training weights	(3, 4)	–

Continued on next page

Table A.1 (continued)

Title	Summary	Cat.	Apps.
TopoBERT: Exploring the topology of fine-tuned word representations (Rathore et al. 2023)	Mapper graph of transformer-based models with applications in finetuning of language models	(3)	–
Experimental observations of the topology of convolutional neural network activations (Purvine et al. 2023)	Analysis of the Mapper graph of neuron activations and definition of a similarity function between layers using sliced Wasserstein distances between persistence diagrams generated from them	(3)	–
ReLU neural networks, polyhedral decompositions, and persistent homology (Liu et al. 2023b)	Detection of homological features in manifolds embedded in the input space of a neural network using the polyhedra decomposition induced by a ReLU feedforward neural network	(2)	–
Caveats of neural persistence in deep neural networks (Girrbach et al. 2023)	Connection between neural persistence and variance measures of neural network weights	(3)	–
On the use of persistent homology to control the generalization capacity of a neural network (Barbara et al. 2024)	Regression of generalization gap using the average persistence of zero dimensional persistence diagrams	(3)	(6)
Visualizing and analyzing the topology of neuron activations in deep adversarial training (Zhou et al. 2023)	Analysis of the Mapper graph of neuron activations of normally and adversarially trained neural networks	(3)	–
TopP&R: Robust support estimation approach for evaluating fidelity and diversity in generative models (Kim et al. 2023)	Approximation of the precision and recall scores for generative models using a persistent homology-based approach	(2)	(7)
Topological structure of complex predictions (Liu et al. 2023a)	Analysis of the GTDA Reeb network of output spaces of neural networks	(2)	–
Topological dynamics of functional neural network graphs during reinforcement learning (Muller et al. 2024)	Study of reinforcement learning neural networks during training and inference using homology	(3)	–
Predicting the generalization gap in neural networks using topological data analysis (Ballester et al. 2024b)	Regression of generalization gap using bootstrapped topological vectorisations of persistence diagrams	(3)	(6)
Unraveling convolution neural networks: A topological exploration of kernel evolution (Yang et al. 2024)	Study of the connection between the Betti curves of the convolution kernels during training and the effectivity of the training process	(3, 4)	–
Functional loops: Monitoring functional organization of deep neural networks using algebraic topology (Zhang and Lin 2024)	Study of the connection between one dimensional persistent homology of neuron activations and their correlations and network performance	(3, 4)	(1)
Topological obstruction to the training of shallow ReLU neural networks (Nurisso et al. 2024)	Study of the topology of the parameter space containing gradient flow trajectories	(4)	–

The Mapper Algorithm

Algorithm 1 Mapper

Require: $\mathcal{D}$ with $|D| = m$, filter function $f : \mathcal{D} \to \mathbb{R}^d$, finite cover $\mathcal{U} = \{\mathcal{U}_i\}_{i \in I}$ of $\text{Im}(f) \subseteq \mathbb{R}^d$, clustering algorithm C.
Ensure: Simplicial complex $S_{\mathcal{D}}$.

1:   $S_{\mathcal{D}} \leftarrow \emptyset$
2:   $\mathcal{D}_i \leftarrow f^{-1}(\mathcal{U}_i)$ for all $i \in I$
3:   **for all** $i \in I$ **do**
4:     $\{C_i^1, \ldots, C_i^{k_i}\} \leftarrow C(\mathcal{D}_i)$ {Apply the clustering algorithm to $\mathcal{D}_i$: the output is a set of clusters}
5:     $S_{\mathcal{D}} \leftarrow S_{\mathcal{D}} \cup \{C_i^1, \ldots, C_i^{k_i}\}$ {Insert the identified clusters as vertices}
6:   **end for**
7:   **for all** $\{C_1, \ldots, C_t\} \in \mathcal{P}\left(\bigcup_{i \in I}\{C_i^1, \ldots, C_i^{k_i}\}\right)$ {For all subsets of the found clusters} **do**
8:     **if** $\bigcap_{j=1}^{t} C_j \neq \emptyset$ **then**
9:      $S_{\mathcal{D}} \leftarrow S_{\mathcal{D}} \cup \{\{C_1, \ldots, C_t\}\}$ {Add the simplex $\{C_1, \ldots, C_t\}$}
10:    **end if**
11: **end for**
12: **return** $S_{\mathcal{D}}$

References

Théo Lacombe, Yuichi Ike, Mathieu Carrière, Frédéric Chazal, Marc Glisse, and Yuhei Umeda. Topological uncertainty: Monitoring trained neural networks through persistence of activation graphs. In Zhi-Hua Zhou, editor, *Proceedings of the Thirtieth International Joint Conference on Artificial Intelligence, IJCAI-21*, pages 2666–2672. International Joint Conferences on Artificial Intelligence Organization, 8 2021. URL https://doi.org/10.24963/ijcai.2021/367. Main Track.

Monica Bianchini and Franco Scarselli. On the complexity of neural network classifiers: A comparison between shallow and deep architectures. *IEEE Transactions on Neural Networks and Learning Systems*, 25(8):1553–1565, 2014. https://doi.org/10.1109/TNNLS.2013.2293637.

Jeremy Charlier, Radu State, et al. PHom-GeM: Persistent homology for generative models. In *The 6th Swiss Conference on Data Science (SDS), 2019 IEEE International Conference*. IEEE, 2019.

Chao Chen, Xiuyan Ni, Qinxun Bai, and Yusu Wang. A topological regularizer for classifiers via persistent homology. In Kamalika Chaudhuri and Masashi Sugiyama, editors, *Proceedings of the Twenty-Second International Conference on Artificial Intelligence and Statistics*, volume 89 of *Proceedings of Machine Learning Research*, pages 2573–2582. PMLR, 16–18 Apr 2019. URL https://proceedings.mlr.press/v89/chen19g.html.

Samir Chowdhury, Thomas Gebhart, Steve Huntsman, and Matvey Yutin. Path homologies of deep feedforward networks. In *2019 18th IEEE International Conference on Machine Learning and Applications (ICMLA)*, pages 1077–1082, 2019. https://doi.org/10.1109/ICMLA.2019.00181.

Valentin Khrulkov and Ivan Oseledets. Geometry score: A method for comparing generative adversarial networks. In Jennifer Dy and Andreas Krause, editors, *Proceedings of the 35th International Conference on Machine Learning*, volume 80 of *Proceedings of Machine Learning Research*, pages 2621–2629. PMLR, 10–15 Jul 2018. URL https://proceedings.mlr.press/v80/khrulkov18a.html.

Pum Jun Kim, Yoojin Jang, Jisu Kim, and Jaejun Yoo. TopP&R: Robust support estimation approach for evaluating fidelity and diversity in generative models. In *Thirty-seventh Conference on Neural Information Processing Systems*, 2023. URL https://openreview.net/forum?id=2gUCMr6fDY.

Weizhi Li, Gautam Dasarathy, Karthikeyan Natesan Ramamurthy, and Visar Berisha. Finding the homology of decision boundaries with active learning. In Hugo Larochelle, Marc'Aurelio Ranzato, Raia Hadsell, Maria-Florina Balcan, and Hsuan-Tien Lin, editors, *Advances in Neural Information Processing Systems*, volume 33, pages 8355–8365. Curran Associates, Inc., 2020. URL https://proceedings.neurips.cc/paper_files/paper/2020/file/5f14615696649541a025d3d0f8e0447f-Paper.pdf.

Meng Liu, Tamal K. Dey, and David F. Gleich. Topological structure of complex predictions. *Nature Machine Intelligence*, Nov 2023a. ISSN 2522-5839. URL https://doi.org/10.1038/s42256-023-00749-8.

Yajing Liu, Christina M. Cole, Chris Peterson, and Michael Kirby. ReLU neural networks, polyhedral decompositions, and persistent homology. In *2nd Annual TAG in Machine Learning*, 2023b.

Giovanni Petri and António Leitão. On the topological expressive power of neural networks. In *TDA & Beyond*, 2020. URL https://openreview.net/forum?id=I44kJPuvqPD.

Karthikeyan Natesan Ramamurthy, Kush Varshney, and Krishnan Mody. Topological data analysis of decision boundaries with application to model selection. In Kamalika Chaudhuri and Ruslan Salakhutdinov, editors, *Proceedings of the 36th International Conference on Machine Learning*, volume 97 of *Proceedings of Machine Learning Research*, pages 5351–5360. PMLR, 09–15 Jun 2019. URL https://proceedings.mlr.press/v97/ramamurthy19a.html.

Naoki Akai, Takatsugu Hirayama, and Hiroshi Murase. Experimental stability analysis of neural networks in classification problems with confidence sets for persistence diagrams. *Neural Networks*, 143:42–51, 2021. ISSN 0893-6080. https://doi.org/10.1016/j.neunet.2021.05.007. URL https://www.sciencedirect.com/science/article/pii/S0893608021001994.

Rubén Ballester, Xavier Arnal Clemente, Carles Casacuberta, Meysam Madadi, Ciprian A. Corneanu, and Sergio Escalera. Predicting the generalization gap in neural networks using topological data analysis. *Neurocomputing*, 596:127787, 2024b. ISSN 0925-2312. https://doi.org/10.1016/j.neucom.2024.127787. URL https://www.sciencedirect.com/science/article/pii/S0925231224005587.

Serguei Barannikov, Ilya Trofimov, Nikita Balabin, and Evgeny Burnaev. Representation topology divergence: A method for comparing neural network representations. In Kamalika Chaudhuri, Stefanie Jegelka, Le Song, Csaba Szepesvari, Gang Niu, and Sivan Sabato, editors, *Proceedings of the 39th International Conference on Machine Learning*, volume 162 of *Proceedings of Machine Learning Research*, pages 1607–1626. PMLR, 17–23 Jul 2022. URL https://proceedings.mlr.press/v162/barannikov22a.html.

Abir Barbara, Younès Bennani, and Joseph Karkazan. On the use of persistent homology to control the generalization capacity of a neural network. In Biao Luo, Long Cheng, Zheng-Guang Wu, Hongyi Li, and Chaojie Li, editors, *Neural Information Processing*, pages 274–286, Singapore, 2024. Springer Nature Singapore. ISBN 978-981-99-8132-8.

Rickard Brüel Gabrielsson and Gunnar Carlsson. Exposition and interpretation of the topology of neural networks. In *2019 18th IEEE International Conference On Machine Learning And Applications (ICMLA)*, pages 1069–1076, 2019. https://doi.org/10.1109/ICMLA.2019.00180.

Gunnar Carlsson and Rickard Brüel Gabrielsson. Topological approaches to deep learning. In Nils A. Baas, Gunnar E. Carlsson, Gereon Quick, Markus Szymik, and Marius Thaule, editors, *Topological Data Analysis*, pages 119–146, Cham, 2020. Springer International Publishing.

Ciprian A. Corneanu, Meysam Madadi, Sergio Escalera, and Aleix M. Martinez. What does it mean to learn in deep networks? And, how does one detect adversarial attacks? In *2019 IEEE/CVF Conference on Computer Vision and Pattern Recognition (CVPR)*, pages 4752–4761, 2019. https://doi.org/10.1109/CVPR.2019.00489.

Ciprian A. Corneanu, Sergio Escalera, and Aleix M. Martinez. Computing the testing error without a testing set. In *2020 IEEE/CVF Conference on Computer Vision and Pattern Recognition (CVPR)*, pages 2674–2682, 2020. https://doi.org/10.1109/CVPR42600.2020.00275.

Thomas Gebhart, Paul Schrater, and Alan Hylton. Characterizing the shape of activation space in deep neural networks. In *2019 18th IEEE International Conference on Machine Learning and Applications (ICMLA)*, pages 1537–1542, 2019. https://doi.org/10.1109/ICMLA.2019.00254.

Morgane Goibert, Elvis Dohmatob, and Thomas Ricatte. An adversarial robustness perspective on the topology of neural networks. In *NeurIPS ML Safety Workshop*, 2022. URL https://openreview.net/forum?id=EtGd7pF237i.

Matthew Muller, Steve Kroon, and Stephan Chalup. Topological dynamics of functional neural network graphs during reinforcement learning. In Biao Luo, Long Cheng, Zheng-Guang Wu, Hongyi Li, and Chaojie Li, editors, *Neural Information Processing*, pages 190–204, Singapore, 2024. Springer Nature Singapore. ISBN 978-981-99-8138-0.

Emilie Purvine, Davis Brown, Brett Jefferson, Cliff Joslyn, Brenda Praggastis, Archit Rathore, Madelyn Shapiro, Bei Wang, and Youjia Zhou. Experimental observations of the topology of convolutional neural network activations. In *Proceedings of the Thirty-Seventh AAAI Conference on Artificial Intelligence and Thirty-Fifth Conference on Innovative Applications of Artificial Intelligence and Thirteenth Symposium on Educational Advances in Artificial Intelligence*, AAAI'23/IAAI'23/EAAI'23. AAAI Press, 2023. ISBN 978-1-57735-880-0. URL https://doi.org/10.1609/aaai.v37i8.26134.

Archit Rathore, Yichu Zhou, Vivek Srikumar, and Bei Wang. TopoBERT: Exploring the topology of fine-tuned word representations. *Information Visualization*, 22(3):186–208, 2023. URL https://doi.org/10.1177/14738716231168671.

Bastian Rieck, Matteo Togninalli, Christian Bock, Michael Moor, Max Horn, Thomas Gumbsch, and Karsten Borgwardt. Neural persistence: A complexity measure for deep neural networks using algebraic topology. In *International Conference on Learning Representations*, 2019. URL https://openreview.net/forum?id=ByxkijC5FQ.

Satoru Watanabe and Hayato Yamana. Deep neural network pruning using persistent homology. In *2020 IEEE Third International Conference on Artificial Intelligence and Knowledge Engineering (AIKE)*, pages 153–156, 2020. https://doi.org/10.1109/AIKE48582.2020.00030.

Satoru Watanabe and Hayato Yamana. Overfitting measurement of convolutional neural networks using trained network weights. *International Journal of Data Science and Analytics*, 14(3): 261–278, Sep 2022a. ISSN 2364-4168. URL https://doi.org/10.1007/s41060-022-00332-1.

Satoru Watanabe and Hayato Yamana. Topological measurement of deep neural networks using persistent homology. *Annals of Mathematics and Artificial Intelligence*, 90(1):75–92, Jan 2022b. ISSN 1573-7470. URL https://doi.org/10.1007/s10472-021-09761-3.

Lei Yang, Mengxue Xu, and Yunan He. Unraveling convolution neural networks: A topological exploration of kernel evolution. *Applied Sciences*, 14(5), 2024. ISSN 2076-3417. URL https://www.mdpi.com/2076-3417/14/5/2197.

Ben Zhang and Hongwei Lin. Functional loops: Monitoring functional organization of deep neural networks using algebraic topology. *Neural Networks*, 174:106239, 2024. ISSN 0893-6080. https://doi.org/10.1016/j.neunet.2024.106239. URL https://www.sciencedirect.com/science/article/pii/S0893608024001631.

Yang Zhao and Hao Zhang. Quantitative performance assessment of CNN units via topological entropy calculation. In *International Conference on Learning Representations*, 2022. URL https://openreview.net/forum?id=xFOyMwWPkz.

Songzhu Zheng, Yikai Zhang, Hubert Wagner, Mayank Goswami, and Chao Chen. Topological detection of trojaned neural networks. In A. Beygelzimer, Y. Dauphin, P. Liang, and J. Wortman Vaughan, editors, *Advances in Neural Information Processing Systems*, 2021. URL https://openreview.net/forum?id=1r2EannVuIA.

Youjia Zhou, Yi Zhou, Jie Ding, and Bei Wang. Visualizing and analyzing the topology of neuron activations in deep adversarial training. In *Topological, Algebraic and Geometric Learning Workshops 2023*, pages 134–145. PMLR, 2023.

Benjamin Dupuis, George Deligiannidis, and Umut Simsekli. Generalization bounds using data-dependent fractal dimensions. In Andreas Krause, Emma Brunskill, Kyunghyun Cho, Barbara Engelhardt, Sivan Sabato, and Jonathan Scarlett, editors, *Proceedings of the 40th International Conference on Machine Learning*, volume 202 of *Proceedings of Machine Learning Research*, pages 8922–8968. PMLR, 23–29 Jul 2023. URL https://proceedings.mlr.press/v202/dupuis23a.html.

Quynh Nguyen. On connected sublevel sets in deep learning. In Kamalika Chaudhuri and Ruslan Salakhutdinov, editors, *Proceedings of the 36th International Conference on Machine Learning*, volume 97 of *Proceedings of Machine Learning Research*, pages 4790–4799. PMLR, 09–15 Jun 2019. URL https://proceedings.mlr.press/v97/nguyen19a.html.

Marco Nurisso, Pierrick Leroy, and Francesco Vaccarino. Topological obstruction to the training of shallow ReLU neural networks. In *The Thirty-eighth Annual Conference on Neural Information Processing Systems*, 2024. URL https://openreview.net/forum?id=3hcn0UxP72.

Tolga Birdal, Aaron Lou, Leonidas J. Guibas, and Umut Simsekli. Intrinsic dimension, persistent homology and generalization in neural networks. In M. Ranzato, A. Beygelzimer, Y. Dauphin, P. S. Liang, and J. Wortman Vaughan, editors, *Advances in Neural Information Processing Systems*, volume 34, pages 6776–6789. Curran Associates, Inc., 2021. URL https://proceedings.neurips.cc/paper_files/paper/2021/file/35a12c43227f217207d4e06ffefe39d3-Paper.pdf.

Maxime Gabella. Topology of learning in feedforward neural networks. *IEEE Transactions on Neural Networks and Learning Systems*, 32(8):3588–3592, 2021. https://doi.org/10.1109/TNNLS.2020.3015790.

Leander Girrbach, Anders Christensen, Ole Winther, Zeynep Akata, and Almut Sophia Koepke. Caveats of neural persistence in deep neural networks. In *2nd Annual TAG in Machine Learning*, 2023.

Gregory Naitzat, Andrey Zhitnikov, and Lek-Heng Lim. Topology of deep neural networks. *J. Mach. Learn. Res.*, 21(1), Jan 2020. ISSN 1532-4435.

Archit Rathore, Nithin Chalapathi, Sourabh Palande, and Bei Wang. TopoAct: Visually exploring the shape of activations in deep learning. *Computer Graphics Forum*, 40(1):382–397, 2021. https://doi.org/10.1111/cgf.14195. URL https://onlinelibrary.wiley.com/doi/abs/10.1111/cgf.14195.

Matthew Wheeler, Jose Bouza, and Peter Bubenik. Activation landscapes as a topological summary of neural network performance. In *2021 IEEE International Conference on Big Data (Big Data)*, pages 3865–3870, Los Alamitos, CA, USA, dec 2021. IEEE Computer Society. URL https://doi.ieeecomputersociety.org/10.1109/BigData52589.2021.9671368.

Sharon Zhou, Eric Zelikman, Fred Lu, Andrew Y. Ng, Gunnar E. Carlsson, and Stefano Ermon. Evaluating the disentanglement of deep generative models through manifold topology. In *International Conference on Learning Representations*, 2021. URL https://openreview.net/forum?id=djwS0m4Ft_A.

Index

© The Author(s), under exclusive license to Springer Nature Switzerland AG 2026
R. Ballester et al., *Topological Data Analysis for Neural Networks*, SpringerBriefs
in Computer Science, https://doi.org/10.1007/978-3-032-08283-1

MIX
Papier aus verantwortungsvollen Quellen
Paper from responsible sources
FSC® C105338
FSC
www.fsc.org

If you have any concerns about our products,
you can contact us on
ProductSafety@springernature.com

In case Publisher is established outside the EU,
the EU authorized representative is:
**Springer Nature Customer Service Center GmbH
Europaplatz 3, 69115 Heidelberg, Germany**

Printed by Libri Plureos GmbH
in Hamburg, Germany